Finite Element Modeling for Materials Engineers Using MATLAB®

Oluleke Oluwole

Finite Element Modeling for Materials Engineers Using MATLAB®

 Springer

Dr. Oluleke Oluwole
Mechanical Engineering Department
University of Ibadan
200005 Ibadan
Nigeria
e-mail: leke_oluwole@yahoo.co.uk; lekeoluwole@gmail.com

Additional material to this book can be downloaded from http://extra.springer.com

ISBN 978-1-4471-6144-8   ISBN 978-0-85729-661-0 (eBook)
DOI 10.1007/978-0-85729-661-0
Springer London Dordrecht Heidelberg New York

Preface

The finite-element method occupies an important place in numerical computation in the applied sciences. It forms the basis for a large chunk of numerical method used in simulation of systems and irregular domains. Its importance today has made it an important subject of study for all engineering students. Clinical treatments of the method itself can be found in many traditional finite element books and this book has not made an attempt to replicate this. This book presents the use of this numerical method to materials engineers using the MathWorks® partial differential equation toolbox^{TM} in an attractive yet potent way in the modeling of many materials processes. By doing this it is hoped that the community is challenged to use this potent, fast and efficient tool in engineering analysis and decision making.

This book gives a background treatment of the Galerkin method in Chaps. 2–4. Topics such as developing weak formulations as prelude to solving the finite element equation, interpolation functions, derivation of elemental equations, assembly and sample solutions were treated in these chapters. Chapter 5 gives an overview on the use of the pdetoolbox. Chapter 6 and 7 give different sample problems and their solutions on heat transfer and elasticity in Materials Engineering. Exercises are given at the end of each example problem. Extra materials containing m-files based on the examples in this book are made available and can be accessed at http://extras.springer.com/ for users convenience and to help in solution of exercises in the book.

September 2008

Dr. Oluleke Oluwole

Contents

ix

For the Reader

Note that all the codes in the sample problems given in this book can be accessed on http://extras.springer.com/. Only some of the codes are presented in the Appendix. There are over 50 pages of codes that can easily be copied and pasted on the MATLAB M-file in order to follow the procedures laid out in this book. To follow the modeling procedure, the reader can actually copy portions of the code to be followed on to a new M-file. A new M-file can be opened by clicking on the sheet icon on the MATLAB worksheet. Run the program by clicking on the Debug menu and by double-clicking save and run on the pull-down menu option. This will enable you to follow the solution procedure in an all-color mode. Other portions can be added as the reader goes along. For example, for the code Diriclet.m listed, follow the geometry construction, copy the codes up to the geometry description, and run. Later on to follow boundary conditions, add the codes for boundary conditions and run again. Then add codes for mesh generation…, then PDE coefficients, etc until you get to solve the PDE. The reader can follow step by step in this way.

Chapter 1
Introduction

The finite-element method is the solution of *variationalformulation* of system governing equations applied over a domain discretized into sub-domains (finite elements) with possibility of different boundary conditions on the discretized surfaces.

Since the aim of most analyses is to find unknown functions which satisfy a known set of differential equations in a domain with all sorts of boundary conditions, the finite element method finds useful application in the solution of these problems. Thus, problems solved by the finite element method are either boundary value problems or initial-value problems or both. The solution of most of these problems by exact methods of analysis is not possible thereby leaving the option of solution by approximation methods. Thus, it becomes imperative to be able to formulate the problem as a *variational* problem and to be able to derive the algebraic equations associated to the variational problem. The least squares, collocation, Rayleigh-Ritz and the Galerkin methods are well-established in solution of many engineering problems and these are well-outlined in many books on the finite element method. This book, however, has focused on the use of the Galerkin method for the solution of the finite-element problems stated in the book. The use of MathWorks® pdetoolbox™ presents a faster and efficient tool as it eliminates code writing and post computational analysis.

O. Oluwole, *Finite Element Modeling for Materials Engineers Using MATLAB®*,
DOI: 10.1007/978-0-85729-661-0_1, © Springer-Verlag London Limited 2011

Chapter 2
The Weak Formulation

2.1 Nodal Finite Elements

The nodal finite method is a *variational formulation* of governing equations applied piecewise over a domain divided into nodal subdivisions. The term *variational* here refers to its modern use which permits its use as equivalent weighted integral to the original problem governing equation (see Sect. 2.4). The principle of solution itself may not necessarily be admissible as a variational principle [1]. Clinical treatments of the classical variational formulation terminology can be assessed in other formal texts and handbooks [1–6].

The basis of the nodal finite element method is the representation of the domain by an assemblage of subdivisions called finite elements. These elements are interconnected at nodes or nodal points. The trial function approximates the distribution of the primary variable across the system of finite elements. Polynomials offer ease of manipulation and are commonly used in the nodal expressions.

2.2 Mesh Elements

One-dimensional (1D) elements are line elements, while 2D elements can be triangular or bilinear elements. Three-dimensional mesh elements are polyhedrals or cuboids [7, 8]. Distorted elements can also be used [9, 10].

2.3 The Finite Element Method Procedure

Some steps are involved in the finite elements analysis. These are

1. Discretization of the domain: It consists of selection of the shape of mesh elements and its construction over the whole domain; numbering of the nodes and elements and the coordinates.

O. Oluwole, *Finite Element Modeling for Materials Engineers Using MATLAB®*,
DOI: 10.1007/978-0-85729-661-0_2, © Springer-Verlag London Limited 2011

2. Selection of interpolation function.
3. Derivation of *variational* formulation of the differential equation for a typical element.
4. Derivation of elements stiffness matrix.
5. Assemblage of global stiffness matrix.
6. Imposition of boundary conditions.
7. Solution of assembled global equation.
8. Representation of results in tabular or graphic form.

2.4 Weak Formulation of Governing Equations

The main approaches of the finite element method are in the redirection of the differential equation of the continuum problem to its integral form and using a trial function over the nodal form of the equation.

Let us take an approximate trial function as $\tilde{u} = \sum_{i=1}^{n} h_i u_i$ where h_i is the set of interpolation functions; u_i is the set of nodal primary variable (displacement, temperature, etc.).

Thus, this function is an approximation solution in the elemental domain defined by a set of integral form of the original differential equation.

Thus, a problem in 3D, defined mathematically by a set of differential equations, D valid in a domain Ω together with the associated boundary condition B can be expressed *in a weak formulation* as

$$\int_v \mathrm{WD}(\tilde{u})\mathrm{d}v + \oint_s \mathrm{WB}(\tilde{u})\mathrm{d}s = 0 \tag{2.1}$$

where w is the weighting function and $\tilde{u}$ is the trial (approximation) function. The first term in Eq. 2.1 is further subjected to integration by parts.

Thus, (2.1) is the approximate form of (2.2)

$$D(u) = 0 \tag{2.2}$$

When $w_i = h_i$, the method is the Galerkin method.

2.5 Gradient and Divergence Theorems

These theorems are used in the derivation of weak formulation. Let A and B be scalar functions defined on a 3D domain.

2.5.1 The Gradient Theorem

$$\int_{\Omega} \mathrm{grad}\,(A)\,\mathrm{d}x\,\mathrm{d}y\,\mathrm{d}z = \int \nabla A\,\mathrm{d}x\,\mathrm{d}y\,\mathrm{d}z = \oint_{\Gamma} \hat{n}A\,\mathrm{d}s$$

$$\int_{\Omega} \left(\hat{i}\frac{\partial A}{\partial x} + \hat{j}\frac{\partial A}{\partial y} + \hat{k}\frac{\partial A}{\partial z} \right)\mathrm{d}x\,\mathrm{d}y\,\mathrm{d}z = \int_{\Gamma} \left(\hat{i}n_x + \hat{j}n_y + \hat{k}n_z \right)A\,\mathrm{d}s$$

where ∇ is the gradient operator $= \hat{i}\dfrac{\partial}{\partial x} + \hat{j}\dfrac{\partial}{\partial y} + \hat{k}\dfrac{\partial}{\partial z}$

2.5.2 Divergence Theorem

$$\int_{\Omega} \mathrm{div}(B)\mathrm{d}x\,\mathrm{d}y\,\mathrm{d}z = \int_{\Omega} \nabla \cdot B\,\mathrm{d}x\,\mathrm{d}y\,\mathrm{d}z = \oint_{\Gamma} \hat{n} \cdot B\,\mathrm{d}s$$

$$\int_{\Omega} \left(\frac{\partial B_x}{\partial x} + \frac{\partial B_y}{\partial y} + \frac{\partial B_z}{\partial z} \right)\mathrm{d}x\,\mathrm{d}y\,\mathrm{d}z = \oint_{\Gamma} \left(n_x B_x + n_y B_y + n_z B_z \right)\mathrm{d}s$$

Subsequently from Sects. 2.5.1 and 2.5.2 we can derive the following which are applicable in the integration by parts of partial differential equations.

1.

$$\int_{\Omega} (\nabla A)B\mathrm{d}x\,\mathrm{d}y\,\mathrm{d}z = - \int_{\Omega} (\nabla B)A\mathrm{d}x\,\mathrm{d}y\,\mathrm{d}z + \oint_{\Gamma} \hat{n}AB\,\mathrm{d}s$$

2.

$$\int (\nabla^2 A)B\mathrm{d}x\,\mathrm{d}y\,\mathrm{d}z = - \int_{\Omega} \nabla A . \nabla B\mathrm{d}x\,\mathrm{d}y\,\mathrm{d}z + \oint_{\Gamma} \frac{\mathrm{d}A}{\mathrm{d}n}B\,\mathrm{d}s$$

where

$$\nabla^2 - \text{laplacian operator} - \frac{\partial^2}{\partial x^2} + \frac{\partial^2}{\partial y^2} + \frac{\partial^2}{\partial z^2}$$

and

$$\nabla = \text{del operator} = n_x\frac{\partial}{\partial x} + n_y\frac{\partial}{\partial y} + n_z\frac{\partial}{\partial z}.$$

These text [11, 12] will be found useful.

2.6 Integration by Parts

If A and B are sufficiently differentiable 1D functions, then the following are applicable:

For a first order differential equation,

$$\int_{x_1}^{x_2} A \left[\frac{\mathrm{d}B}{\mathrm{d}x} \right] \mathrm{d}x = - \int_{x_1}^{x_2} B \frac{\mathrm{d}A}{\mathrm{d}x} \mathrm{d}x + [AB]_{x_1}^{x_2}$$

For a second order differential equation,

$$\int_{x_1}^{x_2} A \left[\frac{\mathrm{d}^2 B}{\mathrm{d}x^2} \right] \mathrm{d}x = - \int_{x_1}^{x_2} \frac{\mathrm{d}B}{\mathrm{d}x} \frac{\mathrm{d}A}{\mathrm{d}x} \mathrm{d}x + \left[A \frac{\mathrm{d}B}{\mathrm{d}x} \right]_{x_1}^{x_2}$$

For a fourth order differential equation,

$$\int_{x_1}^{x_2} A \left[\frac{\mathrm{d}^4 B}{\mathrm{d}x^4} \right] \mathrm{d}x = - \int_{x_1}^{x_2} \frac{\mathrm{d}^2 B}{\mathrm{d}x^2} \frac{\mathrm{d}^2 A}{\mathrm{d}x^2} \mathrm{d}x + \left| \frac{\mathrm{d}A}{\mathrm{d}x} \frac{\mathrm{d}^2 B}{\mathrm{d}x^2} \right|_{x_1}^{x_2} + A \frac{\mathrm{d}^3 B}{\mathrm{d}x^3} \bigg|_{x_1}^{x_2}$$

2.7 Weak Formulations

Strong formulation involves evaluation of the highest order of the derivative term in the differential equation. For an example take a 1D second order differential equation $\mathrm{d}^2 u / \mathrm{d}x^2 = 0$, $0 \le x \le 1$. In the weighted residual method, w being the weighting or test function and $\tilde{u}$ the approximate solution or the trial function when applied to the equation becomes $\int_0^1 w \left[\mathrm{d}^2 \tilde{u} / \mathrm{d}x^2 \right] \mathrm{d}x$. Of course $\mathrm{d}^2 \tilde{u} / \mathrm{d}x^2$ is the residual of the original differential equation $\mathrm{d}^2 u / \mathrm{d}x^2$. The integral must have a non-zero finite value to be an approximate solution to the differential equation.

Thus, there is a problem of finding appropriate approximation (trial) function for a strong formulation which must be differentiable in the order of degree of the given differential equation and at the same time has a non-zero finite value.

This problem is removed when integration by parts is applied to the strong formulation reducing it to a weak formulation.

Thus

$$\int_0^1 w \frac{\mathrm{d}^2 u}{\mathrm{d}x^2} \mathrm{d}x = 0 \text{ becomes } \int_0^1 \left(\frac{\mathrm{d}w}{\mathrm{d}x} \frac{\mathrm{d}u}{\mathrm{d}x} \right) \mathrm{d}x + \left[w \frac{\mathrm{d}u}{\mathrm{d}x} \right]_0^1 = 0$$

Weak formulations applied over sub domains represent the Finite Element equation. Thus, instead of defining trial function in terms of generalized coefficients, the trial function is defined in terms of the nodal variables.

Example 2.1 One-dimensional, second order differential equations
Consider the differential equation

$$\frac{\mathrm{d}}{\mathrm{d}x}\left(a\frac{\mathrm{d}u}{\mathrm{d}x}\right)+f=0 \text{ for } 0<x<1 \tag{2.3}$$

subject to the boundary conditions $u(0)=0$ and $\left(a\frac{\mathrm{d}u(1)}{\mathrm{d}x}\right)=1$.

The *weak* formulation can be obtained through the following steps. Apply integration by parts (see Sect. 2.5.2)

$$0=\int_0^1 w\left[\frac{\mathrm{d}}{\mathrm{d}x}\left(a\frac{\mathrm{d}u}{\mathrm{d}x}\right)+f\right]\mathrm{d}x \tag{2.4}$$

$$0=\int_0^1\left(-a\frac{\mathrm{d}w}{\mathrm{d}x}\frac{\mathrm{d}u}{\mathrm{d}x}+wf\right)\mathrm{d}x+\left[wa\frac{\mathrm{d}u}{\mathrm{d}x}\right]\Big|_0^1 \tag{2.5}$$

Equation 2.5 is the weak formulation.
Applying the boundary conditions would give

$$0=\int_0^1\left(-a\frac{\mathrm{d}w}{\mathrm{d}x}\frac{\mathrm{d}u}{\mathrm{d}x}+wf\right)\mathrm{d}x+w(1) \tag{2.6}$$

Equation 2.6 is the weak formulation with applied boundary condition.
The variational formulation in (2.6) can be expressed as

$$0=B(w,u)-l(w)$$

The quadratic functional $I(u)$ of a variational formulation is represented as

$$I(u)=\frac{1}{2}B(u,u)-l(u)$$

and for the equation under examination this is

$$I(u)=\int_0^1\left(\frac{1}{2}a(\mathrm{d}u/\mathrm{d}x)^2-uf\right)\mathrm{d}x-u(1)$$

The quadratic functional $I(u)$ represents energy in many engineering applications, the minimization of which gives equilibrium solution to the problem. More insight into the use of functionals in finite element analyses can be found in further texts [13–17]. It should be noted though, that in a typical finite element analysis, the domain would be divided into finite elements each having boundary conditions. In such analysis, the weak formulation that would be applied over the elements will be Eq. 2.5 the limits now being the element dimensions. Thus,

$$0 = \sum_{i=1}^{n} \int_{x_i}^{x_{i+1}} \left(-a \frac{dw}{dx} \frac{du}{dx} + wf \right) dx + \left(wa \frac{du}{dx} \right) \Bigg|_{x_i}^{x_{i+1}}$$

represents the weak formulation of the finite element equation over a domain of $n-1$ elements and n nodes.

Example 2.2 Two-dimensional, second order differential equations
Consider the equation

$$k \left(\frac{\partial^2 u}{\partial x^2} + \frac{\partial^2 u}{\partial y^2} \right) = 0 \text{ in } \Omega$$

The weak or variational formulation can be obtained through the following steps:

Weight the integral and obtain the integration by parts of the weighted integral.

$$0 = \int_{\Omega} \left[wk \left(\frac{\partial^2 u}{dx^2} + \frac{\partial^2 u}{\partial y^2} \right) \right] dx \, dy \tag{2.7}$$

$$0 = \int_{\Omega} -k \left[\frac{\partial w}{\partial x} \frac{\partial u}{\partial x} + \frac{\partial w}{\partial y} \frac{\partial u}{\partial y} \right) dx \, dy + \oint_{\Gamma} wk \left(\frac{\partial u}{\partial x} n_x + \frac{\partial u}{\partial y} n_y \right) ds \tag{2.8}$$

Equation 2.8 is the weak formulation to be applied to the system divided into finite elements. This is a typical heat conduction equation. In this case substitute, $u = T$.

If we assume the domain is of rectangular geometry to be solved as a monolithic entity having specified boundary conditions, we can go ahead and substitute the boundary conditions into the weak formulation. Assume the boundary condition is convective on one side in the x-direction; (i.e., $k \partial T / \partial x = -h(T - T_\infty)$) and insulated on the remaining sides (i.e., $\partial T / \partial n = 0$ or $q = 0$).

Then, the boundary integral $\oint w(k \partial T / \partial n) \, ds$ becomes

$$\oint w \left(k \frac{\partial T}{\partial n} \right) ds = \int_{\Gamma_1 + \Gamma_2 + \Gamma_3} w.0 \, ds - \int_{\Gamma_4} w[h(T - T_\infty)] ds = - \int_{\Gamma_4} wh(T - T_\alpha) dx$$

The weak formulation in this case will be:

$$0 = \int_{\Omega} -k \left[\frac{\partial w}{\partial x} \frac{\partial T}{\partial x} + \frac{\partial w}{\partial y} \frac{\partial T}{\partial y} \right) dx \, dy - \int_{\Gamma_4} wh(T - T_\beta) \, dx$$

and the quadratic functional

$$I(T) = \int_{\Omega} \frac{k}{2}\left[\left(\frac{\partial T}{\partial x}\right)^2 + \left(\frac{\partial T}{\partial y}\right)^2\right]\,dx\,dy + \int_{\Gamma_4} h\left(T^2 - TT_\alpha\right)\,dx$$

Example 2.3 Three-dimensional, second order differential equations
Consider the three-dimensional second order PDE.

$$\frac{\partial}{\partial x}\left(k\frac{\partial u}{\partial x}\right) + \frac{\partial}{\partial y}\left(k_2\frac{\partial u}{\partial y}\right) + \frac{\partial}{\partial z}\left(k_3\frac{\partial u}{\partial z}\right) = f \text{ in } \Omega$$

then the weak formulation over an element Ω^e is derived thus:

$$0 = \int_{\Omega^e} w\left[\frac{\partial}{\partial x}\left(k_1\frac{\partial u}{\partial x}\right) + \frac{\partial}{\partial y}\left(k_2\frac{\partial u}{\partial y}\right) + \frac{\partial}{\partial z}\left(k_3\frac{\partial u}{\partial z}\right) - f\right]\,dx\,dy\,dz \qquad (2.9)$$

$$0 = \int_{\Omega^e} -k_1\left(\frac{\partial w}{\partial x}\frac{\partial u}{\partial x} - k_2\frac{\partial w}{\partial y}\frac{\partial u}{\partial y} - k_3\frac{w}{z}\frac{\partial u}{\partial z} - wf\right)\,dx\,dy\,dz$$

$$+ \oint_{\Gamma^e} w\left(k_1\frac{\partial u}{\partial x}n_x + k_2\frac{\partial u}{\partial y}n_y + k_3\frac{\partial u}{\partial z}n_z\right) \qquad (2.10)$$

Equation (2.10) is the weak formulation.

Example 2.4 Transient problems
Transient problems are time dependent or unsteady state problems.
    Let us consider a 1D equation

$$\frac{\partial^2 u}{\partial x^2} = \frac{\partial u}{\partial t}; \quad 0 < x < 1 \qquad (2.11)$$

The weak formulation is derived thus:

$$0 = \int_0^1 \left(w\frac{\partial u}{\partial t} - \frac{\partial^2 u}{\partial x^2}\right)\,dx \qquad (2.12)$$

$$0 = \int_0^1 \left(w\frac{\partial u}{\partial t} + \frac{\partial w}{\partial x}\cdot\frac{\partial u}{\partial x}\right)\,dx - \left[w\frac{\partial u}{\partial x}\right]_{0,t}^{1,t} \qquad (2.13)$$

Equation 2.13 is the weak formulation without applied boundary conditions.

Example 2.5 Fourth order differential equation
In this example, it is necessary to integrate by parts twice to distribute the

derivative equation between the dependent variable u and the test function w (also the weighting function)

$$\frac{d^2}{dx^2}\left[a\frac{d^2u}{dx^2}\right]+f=0$$

for $0 < x < L$ subject to the boundary conditions

$$\left[\frac{d}{dx}\left(a\frac{d^2u}{dx^2}\right)\right]_{x=L}=0 \text{ and } \left(a\frac{d^2u}{dx^2}\right)\Big|_{x=L}=C$$

The weak formulation is derived as follows:

$$0=\int_0^L w\left[\frac{d^2}{dx^2}\left(a\frac{d^2u}{dx^2}\right)+f\right]dx$$

$$0=\int_0^L\left[\left(-\frac{dw}{dx}\right)\frac{d}{dx}\left(a\frac{d^2u}{dx^2}\right)+wf\right]dx+\left[w\frac{d}{dx}\left(a\frac{d^2u}{dx^2}\right)\right]_0^L$$

Integrating further gives

$$0=\int_0^L a\left[\frac{d^2w}{dx^2}\frac{d^2u}{dx^2}+wf\right]dx+\left[w\frac{d}{dx}\left(a\frac{d^2u}{dx^2}\right)-\frac{dw}{dx}a\frac{d^2u}{dx^2}\right]_0^L$$

This is the weak formulation. Of course when applied to the finite element domain, the integral is taken over the element domain as well as the boundary conditions. Specification of u and du/dx in this equation constitutes the Diriclet or essential boundary conditions and the natural boundary or Neumann conditions are satisfied with the specification of $\left[d/dx\left(a\frac{d^2u}{dx^2}\right)\right]$ which is the shear force and $\left(ad^2u/dx^2\right)$ which is the bending moment.

If we apply the boundary conditions, we obtain

$$0=\int_0^l\left[a\frac{d^2w}{dx^2}\frac{d^2u}{dx^2}+wf\right]dx-\frac{dw}{dx}\Big|_{x=L}C$$

A practical example of Example 2.5 is the transverse deflection of a cantilever beam under transverse loading (f) and end moment (M_0). Such an equation will be

$$\frac{d^2}{dx^2}\left[EI\frac{d^2u}{dx^2}\right)+f=0 \text{ subject to}$$

$$u(0) = 0; \quad \frac{\mathrm{d}u}{\mathrm{d}x}(0) = 0; \quad \mathrm{EI}\left.\frac{\mathrm{d}^2 u}{\mathrm{d}x^2}\right|_{x=L} = M_0$$

Here, $a \equiv \mathrm{EI}$.

Therefore

$$B(w, u) = \int_0^L \mathrm{EI}\frac{\mathrm{d}^2 w}{\mathrm{d}x^2} \cdot \frac{\mathrm{d}^2 u}{\mathrm{d}x^2}\,\mathrm{d}x \quad \text{and} \quad l(w) = \int_0^L -wf\mathrm{d}x + M_0\left.\frac{\mathrm{d}w}{\mathrm{d}x}\right|_{X=L}$$

2.8 Exercises

Derive the weak formulations of the following:

1. The 3D heat transfer equation

$$-K\nabla^2 T + \rho\frac{\partial T}{\partial t} = f \text{ in } \Omega$$

2. Beam under loading force, P

$$\frac{\mathrm{d}^2}{\mathrm{d}x^2}\left(\mathrm{EI}\frac{\mathrm{d}^2 u}{\mathrm{d}x^2}\right) + P_0 = 0; \quad 0 < x < 1 \text{ subject to}$$

$$u(0) = 0 \quad u(1) = 0 \quad \mathrm{EI}\left.\frac{\mathrm{d}^2 u}{\mathrm{d}x^2}\right|_{x=1} = 0$$

3. For a transient heat conduction problem

$$\frac{\partial T}{\partial t} - a\frac{\partial}{\partial x}\left(\frac{\partial T}{\partial x}\right) = f; \quad 0 < x < 1 \text{ subject to}$$

$$T(0) = T_0;$$

$$a\frac{\partial T}{\partial x} = -h_c(T - T_\infty) - \hat{q} \quad \text{on } \Gamma$$

4. $\frac{\mathrm{d}^2 u}{\mathrm{d}x^2} = 4; \quad 0 < x < 1 \quad$ subject to $u(0) = 0$ and $u(1) = 0$

5. $\frac{\mathrm{d}^2 u}{\mathrm{d}x^2} + \frac{\mathrm{d}u}{\mathrm{d}x} + u = x \quad 0 < x < 1 \quad$ subject to $u(0) = 0$ and $u(1) = 5$

6. $x^2\frac{\mathrm{d}^2 u}{\mathrm{d}x^2} + x\frac{\mathrm{d}u}{\mathrm{d}x} + 4u = x \quad 1 < x < 5$ subject to $u(1) = 0$ and $u(4) = 8$

References

1. Reddy JN (2006) An introduction to the finite element method, 3rd edn. McGraw-Hill, New York
2. Finlayson BA (1972) The method of weighted residuals and variational principles. Academic Press, New York
3. Washizu K (1975) Variational methods in elasticity and plasticity. Pergamon Press, New York
4. Oden JT, Reddy JN (1983) Variational methods in theoretical mechanics, 2nd edn. Springer, New York
5. Reddy JN (2002) Energy principles and variational methods in applied mechanics, 2nd edn. Wiley, New York
6. Atluri SN (1987) Variational principles. In: Kaedestuncer H, Norrie DH (eds) Finite element handbook. McGraw-Hill, New York
7. Zienkiewicz OC (1977) The finite element method, 3rd edn. McGraw-Hill, New York
8. Zienkiewicz OC, Morgan K (1983) Finite elements and approximations. McGraw-Hill, New York
9. Stasa FL (1985) Applied finite element analysis for engineers. CBS Publishing Japan Ltd, New York
10. Reddy JN (1986) Applied functional analysis and variational methods in engineering. McGraw Hill, New York
11. Schwartz A (1967) Calculus and analytic geometry. Holt, Reinhart and Winston, New York
12. Kaplan W (1973) Advanced calculus. Addison-Wesley, Massachussettes
13. Zienkiewich OC, Taylor RL (1989) The finite element method, Vol. 1. McGraw-Hill, New York
14. Huebner KH (1975) The finite element method for engineers. Wiley, New York
15. Bathe KJ (1982) Finite element procedures in engineering analysis. Prentice Hall, New Jersey
16. Hughes TJR (2000) The finite element method. Prentice Hall, New Jersey
17. Zienkiewich OC, Taylor RL (1991) The finite element method, Vol. 2. McGraw-Hill, New York

Chapter 3
Linear Interpolation Functions

3.1 Parameter Functions and Interpolating Functions

Parameter functions are the nodal parameters being solved for in problem definitions. For example, in heat-transfer problems, the nodal parameter is temperature. Thus, parameter functions are what we refer to as interpolation functions (see Sect. 3.2) or approximation functions in the finite element equation formulation.

These parameter functions need to satisfy two conditions; *compatibility* and *completeness* to ensure convergence as the domain mesh is refined.

Compatibility condition: For C^n continuous problems, the parameter function and its first n derivatives must be continuous between elements. Thus in problems in which C^0 continuity is sufficient, there is no continuity between elements. Examples of C^0 continuous problems are conduction heat transfer where the weak or variational formulation has a first order derivative (see Examples 2.2 and 2.3). C^1 continuous problems are problems that have second order derivatives in the weak formulation such as in beam bending (see Example 2.5 in Chap. 2).

Completeness condition: For C^n continuous problems, the parameter function must be able to give a constant value as well as constant partial derivatives up to the $(n + 1)$th order as the element size decreases to a point. For C^0 continuous functions, the parameter function and its first partial derivative must give constant values as the element size decreases to a point. An example of a parameter function for a C^0 continuous function is Eq. 3.3 (see Sect. 3.3).

If c_2 is zero, $\tilde{u} = c_1$. Also, $\frac{d\tilde{u}}{dx} = c_2$. Therefore, the completeness requirement is satisfied.

3.2 Interpolation, Weighting and Approximation Functions

Interpolation functions when reconstructed with nodal primary variables become approximation functions e.g.,

O. Oluwole, *Finite Element Modeling for Materials Engineers Using MATLAB®*, DOI: 10.1007/978-0-85729-661-0_3, © Springer-Verlag London Limited 2011

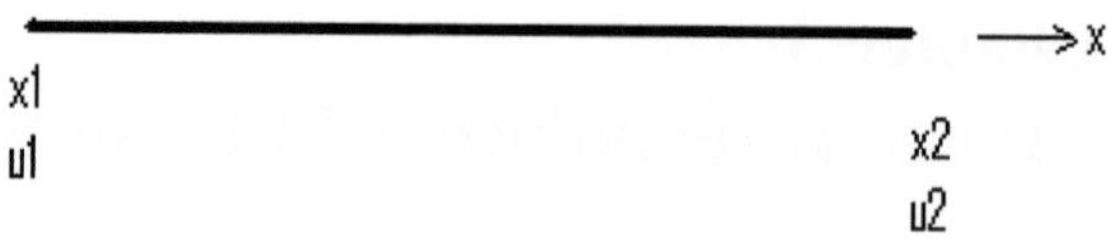

Fig. 3.1 A two-node linear
element

$$\tilde{u} = c_1 + c_2 x = h_1(x)\, u_1 + h_2(x)\, u_2 \qquad (3.1)$$

where $\tilde{u}$ is the interpolation or approximation function u_1 and u_2 are the nodal variables. In the Galerkin method, the weighting function, w becomes $h_1(x)$ and $h_2(x)$ while the approximation function, $\tilde{u}$ is

$$\tilde{u} = h_1(x)\, u_1 + h_2(x)\, u_2 \qquad (3.2)$$

The finite element equation which is now a weak formulation consisting of the weighting function, the approximation function and their derivatives are substituted with the these linear functions and finite values which transform the finite element into a matrix of linearized equations amenable for solution from well-known solution methods. Let us look at examples of linear interpolation functions for one, two and three-dimensional finite element analysis.

3.3 Linear Interpolation Function for One-Dimensional Analysis

One-dimensional problems that are of the C^0 continuous type are represented by two-node linear element as shown in Fig. 3.1 below. The interpolation function, for this problem can be represented by a linear polynomial $\tilde{u}(x)$, where

$$\tilde{u}(x) = c_1 + c_2 x \qquad (3.3)$$

and c_1 and c_2 are constants to be determined from nodal analysis. In matrix form Eq. 3.3 is

$$\tilde{u} = (1 \quad x)\begin{pmatrix} c_1 \\ c_2 \end{pmatrix} \qquad (3.4)$$

The shape function associated with nodes 1 and 2 are shown in Fig. 3.2.

Thus, applying Eq. 3.3 to the two nodes, the linear shape functions associated with the two nodes are

$$\left.\begin{aligned} u(x_1) &= c_1 + c_2 x_1 = u_1 \\ u(x_2) &= c_1 + c_2 x_2 = u_2 \end{aligned}\right\} \qquad (3.5)$$

Placing Eqs. 3.5 in matrix form gives

$$\begin{bmatrix} 1 & x_1 \\ 1 & x_2 \end{bmatrix}\begin{bmatrix} c_1 \\ c_2 \end{bmatrix} = \begin{bmatrix} u_1 \\ u_2 \end{bmatrix} \qquad (3.6)$$

Fig. 3.2 Linear shape function

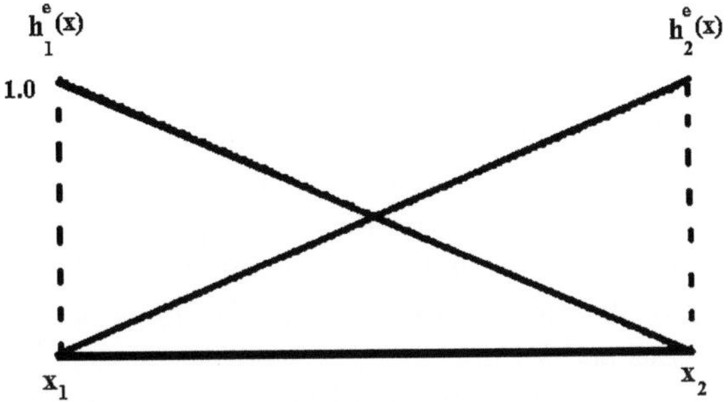

Solving Eq. 3.7 using Cramer's rule gives

$$c_1 = \frac{\begin{vmatrix} u_1 & x_1 \\ u_2 & x_2 \end{vmatrix}}{\begin{vmatrix} 1 & x_1 \\ 1 & x_2 \end{vmatrix}} = \frac{u_1 x_1 - u_2 x_1}{x_2 - x_1} \tag{3.7}$$

$$c_2 = \frac{\begin{vmatrix} 1 & u_1 \\ 1 & u_2 \end{vmatrix}}{\begin{vmatrix} 1 & x_1 \\ 1 & x_2 \end{vmatrix}} = \frac{u_2 - u_1}{x_2 - x_1} \tag{3.8}$$

Placing Eqs. 3.7 and 3.8 back in (3.3) gives

$$\tilde{u} = \frac{u_1 x_2 - u_2 x_1}{x_2 - x_1} + \frac{u_2 - u_1}{x_2 - x_1} \cdot x \tag{3.9}$$

This can be written as

$$\tilde{u} = \frac{u_1 x_2 - u_2 x_1 + u_2 x - u_1 x}{x_2 - x_1} \tag{3.10}$$

Rearranging (3.10) to give form $\tilde{u} = h_1(x)u_1 + h_2(x)u_2$ gives

$$\tilde{u} = \frac{u_1(x_2 - x)}{x_2 - x_1} + \frac{u_2(x - x_1)}{x_2 - x_1} \tag{3.11}$$

Thus,

$$= h_1(x) = \frac{x_2 - x}{x_2 - x_1} = \frac{x_2 - x}{h^e} \tag{3.12}$$

$$h_2(x) = \frac{x - x_1}{x_2 - x_1} = \frac{x - x_1}{h^e} \tag{3.13}$$

$$h^e = x_2 - x_1 = \text{element spacing} \tag{3.14}$$

Equation 3.11 gives an expression for $\tilde{u}$ in terms of the nodal variables and is called the approximation function.

Equations 3.12 and 3.13 are the linear shape functions and are the test or weighting functions in the Galerkin finite element method. You can see the unit value associated with $h_1(x_1) = 1$ and $h_2(x_2) = 1$ (see Fig. 3.2) and derivable from Eqs. 3.12 and 3.13. Also $h_1(x_2) = 0$; and $h_2(x_1) = 0$ using Eqs. 3.12 and 3.13.

Thus, $\tilde{u}(x_1) = u_1$ and $\tilde{u}(x_2) = u_2$ using Eq. 3.11.

Also, the sum of all the shape functions is unity i.e., $\sum_{i=1}^{2} h_i(x) = 1$ from Eqs. 3.12 and 3.13.

Similarly, differentiating (3.12) and (3.13) will give

$$\frac{\mathrm{d}h_1(x)}{\mathrm{d}x} = \frac{0-1}{x_2 - x_1} = \frac{-1}{h^e} \tag{3.15}$$

$$\frac{\mathrm{d}h_2(x)}{\mathrm{d}x} = \frac{1}{x_2 - x_1} = \frac{1}{h^e} \tag{3.16}$$

Differentiating Eq. 3.11 gives;

$$\frac{\mathrm{d}\tilde{u}}{\mathrm{d}x} = \frac{-u_1 + u_2}{x_2 - x_1} = \frac{-u_1 + u_2}{h^e} \tag{3.17}$$

Equations 3.15–3.17 find usefulness during the derivation of element finite element equation which we shall see in Chap. 4. The derivation and use of multiple node elements and higher order functions can be found in other finite element books and handbook [1–4].

3.4 Linear Interpolation Functions for Two-Dimensional Analysis

Discretization in two-dimension domains is done using either the triangular element (Fig. 3.3) or the bilinear rectangular element (Fig. 3.4).

3.4.1 The Linear Triangular Element

The interpolation function for the three-node linear triangular element shown in Fig. 3.3b for C^0 continuous equations is

$$\tilde{u} = c_1 + c_2 x + c_3 y \tag{3.18}$$

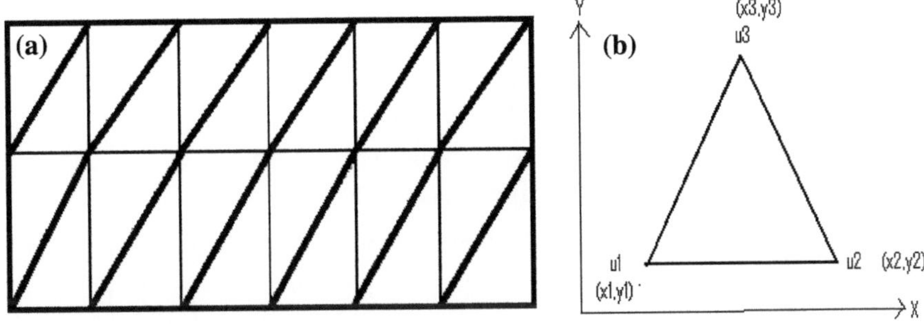

Fig. 3.3 **a** Triangular discretization of two-dimensional domain. **b** The three-noded linear triangular element

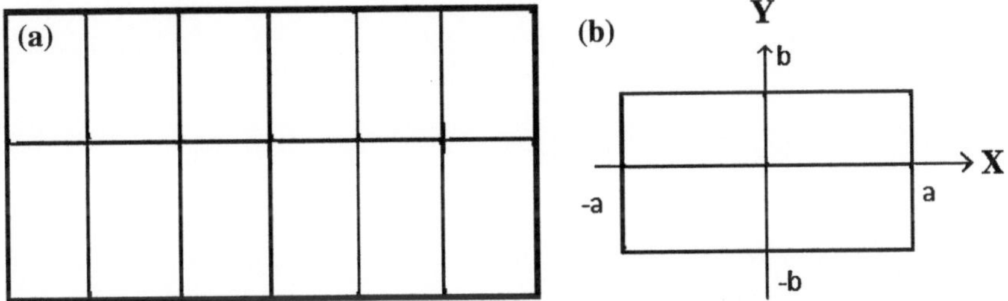

Fig. 3.4 **a** Bilinear discretization of a two-dimensional domain. **b** A four-noded bilinear element

In matrix form (3.18) is

$$\tilde{u} = \begin{pmatrix} 1 & x & y \end{pmatrix} \begin{pmatrix} c_1 \\ c_2 \\ c_3 \end{pmatrix} \tag{3.19}$$

Finding the nodal values at each nodal point gives

$$\begin{pmatrix} u_1 \\ u_2 \\ u_3 \end{pmatrix} = \begin{pmatrix} 1 & x_1 & y_1 \\ 1 & x_2 & y_2 \\ 1 & x_3 & y_3 \end{pmatrix} \begin{pmatrix} c_1 \\ c_1 \\ c_3 \end{pmatrix} \tag{3.20}$$

where x_i and y_i are coordinate values at the ith node. u_1, u_2 and u_3 are the primary nodal variables. Equation 3.20 is of the form $U = BC$.

To find the unknowns (i.e., c_1, c_2 and c_3) we invert the matrix equation. Thus $C = B^{-1}U$

$$\begin{pmatrix} c_1 \\ c_2 \\ c_3 \end{pmatrix} = \frac{1}{2A} \begin{bmatrix} x_2 y_3 - x_3 y_2 & x_3 y_1 & x_1 y_2 - x_2 y_1 \\ y_2 - y_3 & y_3 - y_1 & y_1 - y_2 \\ x_3 - x_2 & x_1 - x_3 & x_2 - x_1 \end{bmatrix} \begin{pmatrix} u_1 \\ u_2 \\ u_3 \end{pmatrix} \tag{3.21}$$

A in matrix Eq. 3.21 is

$$\frac{1}{2}\det\begin{pmatrix} 1 & x_1 & y_1 \\ 1 & x_2 & y_2 \\ 1 & x_3 & y_3 \end{pmatrix} \tag{3.22}$$

which is the area of the linear triangular element.

Substituting (3.22) back in (3.19) gives

$$\tilde{u} = (1 \quad x \quad y)\frac{1}{2A}\begin{pmatrix} x_2y_3 - x_3y_2 & x_3y_1 - x_1y_3 & x_1y_2 - x_2y_1 \\ y_2 - y_3 & y_3 - y_1 & y_1 - y_2 \\ x_3 - x_2 & x_1 - x_3 & x_2 - x_1 \end{pmatrix}\begin{pmatrix} u_1 \\ u_2 \\ u_3 \end{pmatrix}$$

$$\tilde{u} = \frac{1}{2A}\begin{pmatrix} x_2y_3 - x_3y_2 & x_3y_1 - x_1y_3 & x_1y_2 - x_2y_1 \\ x(y_2 - y_3) & x(y_3 - y_1) & x(y_1 - y_2) \\ y(x_3 - x_2) & y(x_1 - x_3) & y(x_2 - x_1) \end{pmatrix}\begin{pmatrix} u_1 \\ u_2 \\ u_3 \end{pmatrix} \tag{3.23}$$

$$\tilde{u} = \frac{1}{2A}[x_2y_3 - x_3y_2 + (y_2 - y_3)x + (x_3 - x_2)y]u_1$$

$$+ \frac{1}{2A}[(x_3y_1 - x_1y_3) + (y_3 - y_1)x + (x_1 - x_3)y]u_2$$

$$+ \frac{1}{2A}[(x_1y_2 - x_2y_1) + (y_1 - y_2)x + (x_2 - x_1)y]u_3 \tag{3.24}$$

Equation 3.24 is of the form

$$\tilde{u} = h_1(x,y)u_1 + h_2(x,y)u_2 + h_3(x,y)u_3 \tag{3.25}$$

Thus,

$$h_1 = \frac{1}{2A}[x_2y_3 - x_3y_2 + (y_2 - y_3)x + (x_3 - x_2)y] \tag{3.26}$$

$$h_2 = \frac{1}{2A}[x_3y_1 - x_1xy + (y_3 - y_1)x + (x_1 - x_3)y] \tag{3.27}$$

$$h_3 = \frac{1}{2A}[x_1y_2 - x_2y_1 + (y_1 - y_2)x + (x_2 - x_1)y] \tag{3.28}$$

Equation 3.24 is the approximation or trial function while Eqs. 3.25–3.27 are the test or weighing functions in the Galerkin finite element method.

$$A = \frac{1}{2}(x_2y_3 + x_1y_2 + y_1x_3 - x_2y_1 - x_3y_2 - y_3x_1) \tag{3.29}$$

$$= \frac{1}{2}[(y_3 - y_1)x_2 - (y_3 - y_1)x_1 - (y_2 - y_1)x_3] \tag{3.30}$$

From Fig. 3.3b; $y_2 - y_1 = 0$ and $y_3 - y_1 = y_3 - y_2$. Therefore

$$A = \frac{1}{2}[(y_3 - y_1)x_2 - (y_3 - y_1)x_1] \tag{3.31}$$

$$= \frac{1}{2}(y_3 - y_1)(x_2 - x_1) \tag{3.32}$$

$= \frac{1}{2}$ base $\times$ height $=$ area of the triangular sub-domain

$$\frac{d\tilde{u}}{dx} = \frac{1}{2A}[(y_2 - y_3)u_1 + (y_3 - y_1)u_2 + (y_1 - y_2)u_3]$$

$$= \frac{1}{2A}[y_2 - y_3 \quad y_3 - y_1 \quad y_1 - y_2] \begin{bmatrix} u_1 \\ u_2 \\ u_3 \end{bmatrix} \tag{3.33}$$

$$\frac{d\tilde{u}}{dy} = \frac{1}{2A}[x_3 - x_2 \quad x_1 - x_3 \quad x_2 - x_1] \begin{bmatrix} u_1 \\ u_2 \\ u_3 \end{bmatrix} \tag{3.34}$$

$$\frac{dh_1}{dx} = \frac{1}{2A}[y_2 - y_3]$$

$$\frac{dh_2}{dx} = \frac{1}{2A}[y_3 - y_1] \tag{3.35}$$

$$\frac{dh_3}{dx} = \frac{1}{2A}[y_1 - y_2]$$

$$\frac{dh_1}{dy} = \frac{1}{2A}[x_3 - x_2]$$

$$\frac{dh_2}{dy} = \frac{1}{2A}[x_1 - x_3] \tag{3.36}$$

$$\frac{dh_3}{dy} = \frac{1}{2A}[x_2 - x_1]$$

Note that Eqs. 3.32–3.36 will be needed in the finite element equation derivation (see Chap. 5).

3.4.2 The Bilinear Element

The interpolation function for the bilinear rectangular element as shown in Fig. 3.2 is

$$\tilde{u} = c_1 + c_2 x + c_3 y + c_4 xy \tag{3.37}$$

$$\tilde{u} = [1 \quad x \quad y \quad xy] \begin{Bmatrix} c_1 \\ c_2 \\ c_3 \\ c_4 \end{Bmatrix} \tag{3.38}$$

Expressing $\tilde{u}$ in form of the nodal values gives

$$\left\{\begin{array}{c} u_1 \\ u_2 \\ u_3 \\ u_4 \end{array}\right\} = \left[\begin{array}{cccc} 1 & x_1 & y_1 & x_1 y_1 \\ 1 & x_2 & y_2 & x_2 y_2 \\ 1 & x_3 & y_3 & x_3 y_3 \\ 1 & x_4 & y_4 & x_4 y_4 \end{array}\right] \left\{\begin{array}{c} c_1 \\ c_2 \\ c_3 \\ c_4 \end{array}\right\} \tag{3.39}$$

Applying the same steps as used in Sect. 3.4.1 gives the shape functions

$$h_1(x,y) = \frac{1}{4ab}(a-x)(b-y)$$

$$h_2(x,y) = \frac{1}{4ab}(a+x)(b-y)$$

$$h_3(x,y) = \frac{1}{4ab}(a+x)(b+y) \tag{3.40}$$

$$h_4(x,y) = \frac{1}{4ab}(a-x)(b+y)$$

where a and b are the element parameters. Thus,

$$\tilde{u} = h_1(x,y)u_1 + h_2(x,y)u_2 + h_3(x,y)u_3 + h_4(x,y)u_4 \tag{3.41}$$

Remember, in the finite element equation, there will always be the need to substitute for du/dx, du/dy, dh/dx and dh/dy where $i = 1, 2, \ldots, 4$ (the element nodal numbers). Thus,

$$\frac{dh_1}{dx}(x,y) = \frac{-1}{4ab}(b-y)$$

$$\frac{dh_1(x,y)}{dy} = \frac{-1(a-x)}{4ab}$$

$$\frac{dh_2(x,y)}{dx} = \frac{1(b-y)}{4ab}$$

$$\frac{dh_2(x,y)}{dy} = \frac{-1(a+x)}{4ab}$$

$$\frac{dh_3(x,y)}{dx} = \frac{1}{4ab}(b+y) \tag{3.42}$$

$$\frac{dh_3(x,y)}{dy} = \frac{1}{4ab}(a+x)$$

$$\frac{dh_4(x,y)}{dx} = \frac{1}{4ab}(b+y)$$

$$\frac{dh_4(x,y)}{dy} = \frac{1}{4ab}(a-x)$$

Also,

$$\frac{d\tilde{u}}{dx} = \frac{dh_1(x,y)u_1}{dx} + \frac{dh_2(x,y)u_2}{dx} + \frac{dh_3(x,y)u_3}{dx} + \frac{dh_4(x,y)u_4}{dx}$$

$$= \begin{bmatrix} \dfrac{dh_1(x,y)}{dx} & \dfrac{dh_2(x,y)}{dx} & \dfrac{dh_3(x,y)}{dx} & \dfrac{dh_4(x,y)}{dx} \end{bmatrix} \begin{Bmatrix} u_1 \\ u_2 \\ u_3 \\ u_4 \end{Bmatrix} \tag{3.43}$$

Similarly, for

$$\frac{d\bar{u}}{dy} = \begin{bmatrix} \dfrac{dh_1(x,y)}{dy} & \dfrac{dh_2(x,y)}{dy} & \dfrac{dh_3(x,y)}{dy} & \dfrac{dh_4(x,y)}{dy} \end{bmatrix} \begin{Bmatrix} u_1 \\ u_2 \\ u_3 \\ u_4 \end{Bmatrix} \tag{3.44}$$

3.5 Linear Interpolation Functions for Three-Dimensional Problems

For interpolation in three-dimensions, four-node tetrahedral elements (Fig. 3.5) or eight-node prism elements (Fig. 3.6) can be used.

3.5.1 Four-Node Tetrahedral Elements

The linear interpolation function can be expressed once again as

$$\tilde{u} = c_1 + c_2 x + c_3 y + c_4 z \tag{3.45}$$

This can be written in terms of local nodal values as

$$\begin{Bmatrix} u_1 \\ u_2 \\ u_3 \\ u_4 \end{Bmatrix} = \begin{bmatrix} 1 & x_1 & y_1 & z_1 \\ 1 & x_2 & y_2 & z_2 \\ 1 & x_3 & y_3 & z_3 \\ 1 & x_4 & y_4 & z_4 \end{bmatrix} \begin{bmatrix} c_1 \\ c_2 \\ c_3 \\ c_4 \end{bmatrix} \tag{3.46}$$

Solving the equations by applying the same techniques as used in Sect. 3.4.1 gives the function

$$\tilde{u} = h_1(x,y,z)u_1 + h_2(x,y,z)u_2 + h_3(x,y,z)u_3 + h_4(x,y,z)u_4 \tag{3.47}$$

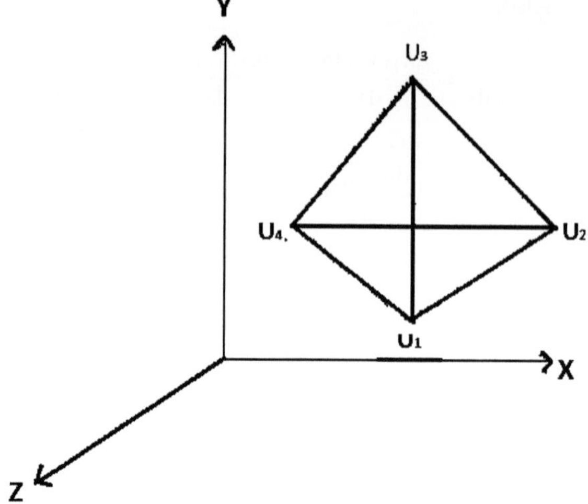

Fig. 3.5 A four-node tetrahedral element (2-D triangular surface elements)

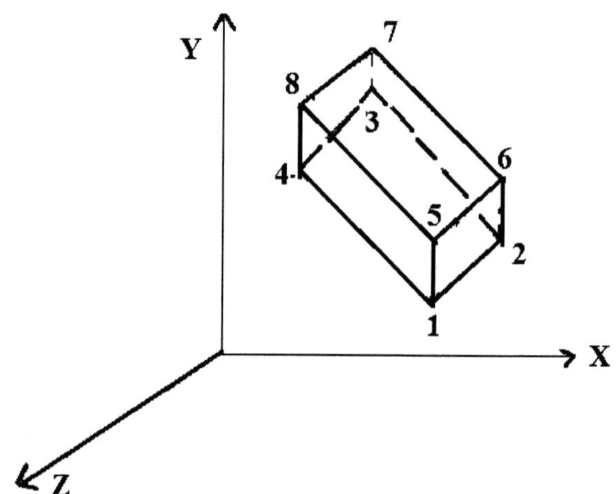

Fig. 3.6 An eight-node brick element (2-D rectangular surface elements)

The shape functions are

$$h_i(x, y, z); \quad i = 1 - 4$$

$$
\begin{aligned}
h_1(x, y, z) &= s_{11} + s_{21}x + s_{31}y + s_{41}z \\
h_2(x, y, z) &= s_{12} + s_{22}x + s_{32}y + s_{42}z \\
h_3(x, y, z) &= s_{13} + s_{23}x + s_{33}y + s_{43}z \\
h_4(x, y, z) &= s_{14} + s_{24}x + s_{34}y + s_{44}z
\end{aligned}
\tag{3.48}
$$

where

$$s_{11} = \frac{1}{6V} \det \begin{bmatrix} x_2 & y_2 & z_2 \\ x_3 & y_3 & z_3 \\ x_4 & y_4 & z_4 \end{bmatrix} \quad \text{and} \quad V = \frac{1}{6} \det \begin{bmatrix} 1 & x_1 & y_1 & z_1 \\ 1 & x_2 & y_2 & z_2 \\ 1 & x_3 & y_3 & z_3 \\ 1 & x_4 & y_4 & z_4 \end{bmatrix} \tag{3.49}$$

3.5.2 Eight-Node Brick Elements

The linear interpolation function can be expressed once again as

$$\tilde{u} = c_1 + c_2 x + c_3 y + c_4 z + c_5 xy + c_6 xz + c_7 yz + c_8 xyz \tag{3.50}$$

The approximation function can be obtained using the same methods as Sect. 3.4.1 The shape functions become

$$h_1(x,y,z) = \frac{1}{8abc}(1-x)(1-y)(1-z)$$
$$h_2(x,y,z) = \frac{1}{8abc}(1+x)(1-y)(1+z)$$
$$h_3(x,y,z) = \frac{1}{8abc}(1+x)(1+y)(1+z)$$
$$h_4(x,y,z) = \frac{1}{8abc}(1-x)(1+y)(1+z)$$
$$h_5(x,y,z) = \frac{1}{8abc}(1-x)(1-y)(1+z)$$
$$h_6(x,y,z) = \frac{1}{8abc}(1+x)(1-y)(1-z)$$
$$h_7(x,y,z) = \frac{1}{8abc}(1+x)(1+y)(1-z)$$
$$h_8(x,y,z) = \frac{1}{8abc}(1-x)(1+y)(1-z)$$

$$\tag{3.51}$$

3.6 Other Coordinate Systems Used in Derivation of Shape Functions

3.6.1 Serendipity Coordinates

The shape functions defined above have been defined in terms of the global coordinates x, y and z. However, Serendipity coordinates can be used as well. These are local, normalized coordinates defined relative to the global coordinate system so named because of the chance discovery of the shape functions for four-node rectangular elements. It has become very convenient to use this coordinate with Gauss–Legendre quadrature is applied [2, 3].

Fig. 3.7 Serendipity
coordinates in a linear
element

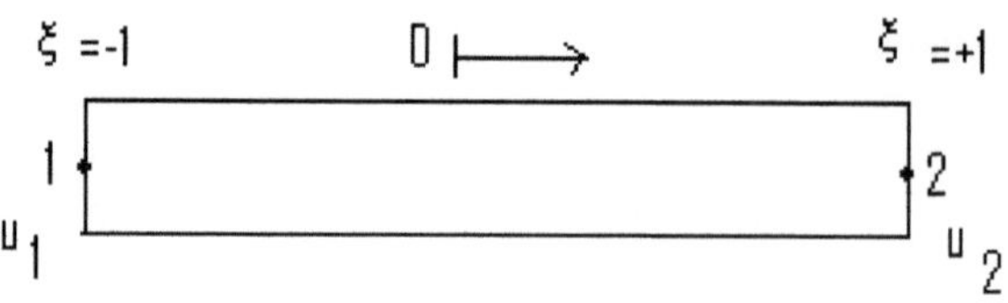

3.6.1.1 Serendipity Coordinates Applied to One-Dimensional Problems

A two-noded lineal element has a serendipity coordinate, ξ (Fig. 3.7) such that
$\xi = 0$ at $\bar{x} = \left(\dfrac{x_1 + x_2}{2}\right)$ and $-1 \le \xi \le 1$

Thus,

$$\xi = \frac{2(x - \bar{x})}{x_2 - x_1} \tag{3.52}$$

Thus the interpolation function becomes $\tilde{u} = h_1(\xi)u_1 + h_2(\xi)u_2$ and the geometric mapping function becomes $x = h_1(\xi)x_1 + h_2(\xi)x_2$. Using this coordinate,
from Eq. 3.52, the shape functions become

$$h_1 = \tfrac{1}{2}(1 - \xi) \quad \text{and} \quad h_2 = \tfrac{1}{2}(1+\xi) \tag{3.53}$$

Note also that this can be derived from the shape functions using global coordinate.

3.6.1.2 Serendipity Coordinates Applied to Two-Dimensional Problems

This coordinate system applied to the bi-lineal element is easily applied like the
lineal element. This time, the coordinates are ξ and η (Fig. 3.8).

$$\xi = 0 \text{ at } \bar{x} = \left(\frac{x_1 + x_2}{2}\right) \text{ and } \eta = 0 \text{ at } \bar{y} = \left(\frac{y_1 + y_2}{2}\right)$$

$$-1 \le \xi \le 1 \text{ and } -1 \le \eta \le 1$$

$$\xi = \frac{(x - \bar{x})}{a} \text{ and } \eta = \frac{(y - \bar{y})}{b} \tag{3.54}$$

Note that $2a = x_1 - x_2$ and $2b = y_1 - y_2$. When these are substituted into the
one-dimensional relation in Eq. 3.50 it gives Eq. 3.54.

Thus, the shape functions which are also the weighting functions in the
Galerkin finite element method become

$$h_1(\xi, \eta) = \frac{1}{4}(1 - \xi)(1 - \eta)$$

$$h_2(\xi, \eta) = \frac{1}{4}(1 + \xi)(1 - \eta)$$

$$h_3(\xi, \eta) = \frac{1}{4}(1 + \xi)(1 + \eta) \tag{3.55}$$

$$h_4(\xi, \eta) = \frac{1}{4}(1 - \xi)(1 + \eta)$$

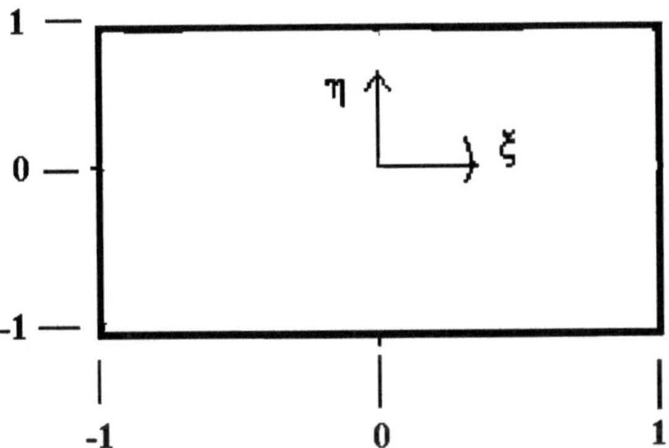

Fig. 3.8 Serendipity coordinates in a rectangular element

Note also that this can be derived from the shape functions using global coordinate.

3.6.1.3 Serendipity Coordinates Applied to Three-Dimensional Problems

Serendipity coordinates can be applied to the brick element in three-dimensional shape function derivation as follows: The coordinates are ξ, η and ζ.

$$\bar{x} = \left(\frac{x_1 + x_2}{2}\right), \quad \bar{y} = \left(\frac{y_1 + y_2}{2}\right) \quad \text{and} \quad \bar{z} = \left(\frac{z_1 + z_2}{2}\right)$$

$$-1 \leq \xi \leq +1, \quad -1 \leq \eta \leq +1 \quad \text{and} \quad -1 \leq \zeta \leq +1$$

$$\xi = \frac{(x - \bar{x})}{a}, \quad \eta = \frac{(y - \bar{y})}{b} \quad \text{and} \quad \zeta = \frac{(z - \bar{z})}{c}$$

element half lengths in the x, y and z directions are 1. This implies that

$$\xi, \eta, \zeta = (0, 0, 0) \text{ at } (\bar{x}, \bar{y}, \bar{z})$$

The shape functions for C^0 continuous problems are:

$$h_1(\xi, \eta, \zeta) = \frac{1}{8}(1 - \xi)(1 - \eta)(1 - \zeta)$$

$$h_2(\xi, \eta, \zeta) = \frac{1}{8}(1 + \xi)(1 - \eta)(1 + \zeta)$$

$$h_3(\xi, \eta, \zeta) = \frac{1}{8}(1 + \xi)(1 + \eta)(1 + \zeta)$$

$$h_4(\xi, \eta, \zeta) = \frac{1}{8}(1 - \xi)(1 + \eta)(1 + \zeta)$$

$$h_5(\xi, \eta, \zeta) = \frac{1}{8}(1 - \xi)(1 - \eta)(1 + \zeta)$$

$$h_6(\xi, \eta, \zeta) = \frac{1}{8}(1 + \xi)(1 - \eta)(1 - \zeta)$$

$$h_7(\xi, \eta, \zeta) = \frac{1}{8}(1 + \xi)(1 + \eta)(1 - \zeta)$$

$$h_8(\xi, \eta, \zeta) = \frac{1}{8}(1 - \xi)(1 + \eta)(1 - \zeta)$$

$$\text{(3.56)}$$

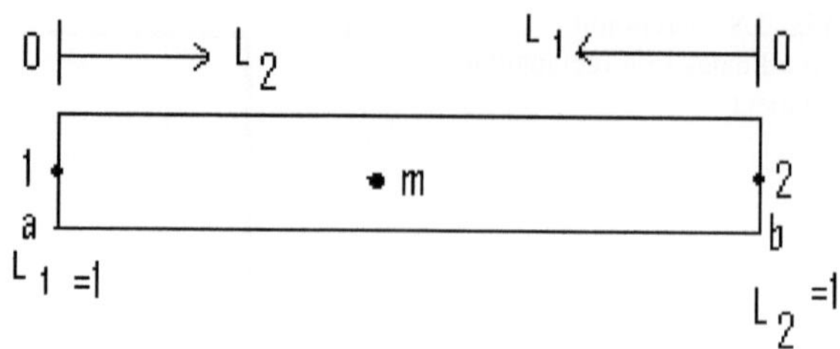

Fig. 3.9 Length coordinates of a two-node lineal element

3.6.2 Length Coordinates

Length coordinates are used for one-dimensional elements. The formulation is such that simple integration formulae can be applied to the integrals thereby easing calculation (see Chap. 5). Length coordinates are also local normalized coordinates like the serendipity coordinates. In this coordinate system, two independent length coordinates, L_1 and L_2 are introduced, each traversing the lineal length from opposite directions to any designated point, m along the lineal length (Fig. 3.9).

Thus, taking $0 \leq L_1 \leq 1$ and $0 \leq L_2 \leq 1$ and $L_1 + L_2 = 1$; $L_2 = 1 - L_1$;

It can be seen that length $L_1 = \dfrac{\text{length } bm}{\text{length } ab}$

Similarly, length $L_2 = \dfrac{\text{length } am}{\text{length } ab}$

Thus, the shape functions will become

$$h_1(L) = L_1 \quad \text{and} \quad h_2(L) = L_2 \tag{3.57}$$

$\tilde{u} = h_1(L)u_1 + h_2(L)u_2$. This implies that the transformation equation is

$$L_1(x) = \frac{x_2 - x}{x_2 - x_1} = \frac{x_2 - x}{l}$$

Mapping from L coordinates to x gives $X = X_1 L_1 + X_2 L_2$

$$\frac{\partial \tilde{u}}{\partial L_1} = \frac{\partial h_1}{\partial x}\frac{\partial x}{\partial L_1}u_1 + \frac{\partial h_2}{\partial x}\frac{\partial x}{\partial L_1}u_2$$

$\dfrac{\partial x}{\partial L_1}$ is the Jacobian term. Since, the length coordinates are natural coordinates, method of solution follows the same procedure outlined under triangular iso-parametric elements (see Sect. 3.7.2).

3.6.3 Area Coordinates

This coordinate system is also a natural coordinate system and is normalized as well. An example of its application to a three-noded triangular element is as shown in Fig. 3.10.

The coordinates here are area coordinates L_1, L_2 and L_3 defined as

$$\text{Area } L_1 = \frac{\text{area } mbc}{\text{area } abc}$$
$$\text{Area } L_2 = \frac{\text{area } mac}{\text{area } abc}$$
$$\text{Area } L_3 = \frac{\text{area } mab}{\text{area } abc}$$

They are related thus;

$$L_1 + L_2 + L_3 = 1$$
$$0 \leq L_1 \leq 1, \quad 0 \leq L_2 \leq 1 \quad \text{and} \quad 0 \leq L_3 \leq 1$$

Note also from Fig. 3.8 that area of Δmac = area of Δsac. Thus, the shape functions for nodes 1, 2, 3 are

$$h_1(L) = L_1, \ h_2(L) = L_2 \text{ and } h_3(L) = L_3 \tag{3.58}$$

The derivation of the derivatives of the shape functions are outlined under Sect. 3.7.2.

3.6.4 Volume Coordinates

Volume coordinates are used for tetrahedral elements. In this case, the coordinates are L_1, L_2, L_3, and L_4.
   Thus,
$$L_i = \frac{\text{Volume } V_i}{\text{Total volume}}, \quad i = 1,2,3,4$$

These volume fractions are related thus; $L_1 + L_2 + L_3 + L_4 = 1$.
The shape functions are

$$h_1(L) = L_1, \ h_2(L) = L_2, \ h_3(L) = L_3 \quad \text{and} \quad h_4(L) = L_4. \tag{3.59}$$

3.7 Isoparametric Elements

Isoparametric elements are elements defined in what is called the 'natural' coordinates system as opposed to global (xyz) coordinate system and at the same time have the nodes used to define the geometry or system at the same location and the same number as the parameter functions sought. Since they are situated on normalized coordinates, it makes it easy to work with deformed elements on complex geometries because they can easily be mapped on to master elements. Thus, the serendipity, area and volume coordinates are used for isometric elements coupled with the Gauss–Legendre integration procedure to facilitate solution of the integral.

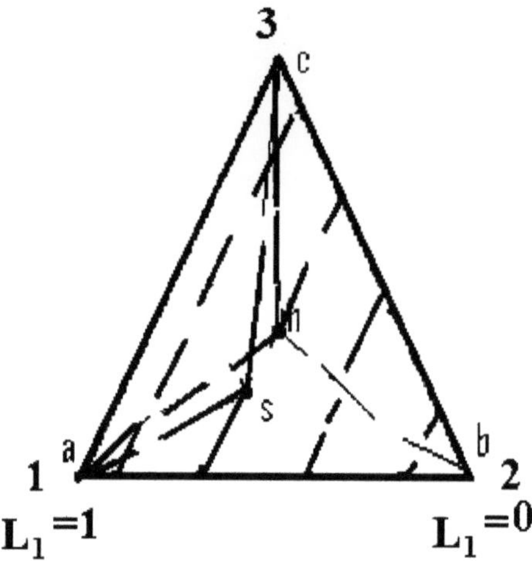

Fig. 3.10 Triangular element showing area coordinates and lines of constant area coordinates

Elements where the parametric nodes are more than the geometrie nodes can also be used in finite element computations. These elements are called *subparametric* elements. In this situation, their interpolation functions are no longer linear [2, 3].

3.7.1 Linear Isoparametric Element

For a linear element (i.e., 2 node element), the $\xi\eta\zeta$ coordinate system is chosen in such a way that the nodes located at ξ axis are located at $\xi_1 = -1.0$ and $\xi_2 = 1.0$ as opposed to the nodes on x- axis (see serendipity element for one-dimensional problem in Sect. 3.6.1.1). The shape functions are:

$$h_1(\xi) = \frac{1}{2}(1 - \xi) \quad \text{and} \quad h_2(\xi) = \frac{1}{2}(1 + \xi) \tag{3.60}$$

This shape functions can be used for the geometric mapping as well as interpolating the parameter u within the element.

The interpolation function becomes,

$$\tilde{u} = h_1(\xi)u_1 + h_2(\xi)u_2 \tag{3.61}$$

and the geometric mapping function;

$$x = h_1(\xi)x_1 + h_2(\xi)x_2 \tag{3.62}$$

To compute values needed in computing element matrices, the chain rule of differentiation can be used as shown below:

$$\begin{aligned}
\frac{\partial \tilde{u}}{\partial \xi} &= \frac{\partial h_1(\xi)}{\partial x}\frac{\partial x}{\partial \xi}u_1 + \frac{\partial h_2(\xi)}{\partial x}\frac{\partial x}{\partial \xi}.u_2 \\
\frac{\partial \tilde{u}}{\partial x} &= \frac{\partial h_1(\xi)}{\partial \xi}\frac{\partial \xi}{\partial x}u_1 + \frac{\partial h_2(\xi)}{\partial \xi}\frac{\partial \xi}{\partial x}.u_2
\end{aligned} \tag{3.63}$$

$J = \dfrac{\partial x}{\partial \xi}$ is called the Jacobian; obtainable from Eq. 3.62

$$\frac{dx}{d\xi} = \frac{dh_1(\xi)}{d\xi}.x_1 + \frac{dh_2(\xi)}{d\xi}x_2 = \frac{1}{2}(x_2 - x_1) = \frac{h_i}{2} \tag{3.64}$$

Substituting (3.64) in (3.63) gives

$$\frac{d\tilde{u}}{dx} - \frac{1}{x_2 - x_1}u_1 + \frac{1}{x_2 - x_1}u_2 = \frac{-1}{h_i}u_1 + \frac{1}{h_i}u_2 \tag{3.65}$$

Also

$$\frac{dh_1(\xi)}{dx} = \frac{1}{J} \cdot \frac{dh_1(\xi)}{d\xi} = \frac{2}{x_2 - x_1} \cdot \frac{1}{2} = -\frac{1}{h_i}$$

and

$$\frac{dh_2(\xi)}{dx} = \frac{1}{x_2 - x_1} = \frac{1}{h_i} \tag{3.66}$$

We can see that the values of $\dfrac{d\tilde{u}}{dx}$, $\dfrac{dh_1(\xi)}{dx}$ and $\dfrac{dh_2(\xi)}{dx}$ for linear isoparametric is the same for the linear elements using physical coordinates (3.12) and (3.13).

Thus, in computing for the weak formulation (see Sect. 3.4), we will have

$$\int_{x_i}^{x_i+1} \left[\frac{dw}{dx}\frac{du}{dx} - wf\right] dx \text{ now becoming } \int_{-1}^{1} \left[\frac{dw}{dx}\frac{du}{dx} - wf\right] J d\xi \tag{3.67}$$

Solving this gives the same equation as Eq. 3.8.

3.7.2 Triangular Isoparametric Element

Area coordinates, L_1, L_2 and L_3 are commonly used for triangular isoparametric elements, where $L_1 + L_2 + L_3 = 1$, L_1 and L_2 are independent and L_3 is dependent. Thus,

$$L_3 = 1 - L_1 - L_2$$
$$L_2 = L_2(x, y)$$
$$L_1 = L_1(x, y); \quad \text{and} \quad 0 \leq L_1 \leq 1, \ 0 \leq L_2 \leq 1 \text{ and } 0 \leq L_3 \leq 1$$

$$\begin{bmatrix} \dfrac{\partial h_i}{\partial L_1} \\[2ex] \dfrac{\partial h_i}{\partial L_2} \end{bmatrix} = \begin{bmatrix} \dfrac{\partial x}{\partial L_1} & \dfrac{\partial y}{\partial L_1} \\[2ex] \dfrac{\partial x}{\partial L_2} & \dfrac{\partial y}{\partial L_2} \end{bmatrix} \begin{bmatrix} \dfrac{\partial h_i}{\partial x} \\[2ex] \dfrac{\partial h_i}{\partial y} \end{bmatrix}$$

The Jacobian in this case is

$$[J] = \begin{bmatrix} \dfrac{\partial x}{\partial L_1} & \dfrac{\partial y}{\partial L_1} \\[2mm] \dfrac{\partial x}{\partial L_2} & \dfrac{\partial y}{\partial L_2} \end{bmatrix} \tag{3.68}$$

$$\begin{bmatrix} \dfrac{\partial h_i}{\partial x} \\[2mm] \dfrac{\partial h_i}{\partial y} \end{bmatrix} = \begin{bmatrix} \dfrac{\partial x}{\partial L_1} & \dfrac{\partial y}{\partial L_1} \\[2mm] \dfrac{\partial x}{\partial L_2} & \dfrac{\partial y}{\partial L_2} \end{bmatrix}^{-1} \begin{bmatrix} \dfrac{\partial h_i}{\partial L} \\[2mm] \dfrac{\partial h_i}{\partial L} \end{bmatrix}$$

Alternatively, using the partial differentiation route,

$$\frac{\partial h_i}{\partial L_1} = \left(\frac{\partial h_1}{\partial L_1}\right)_{L_2 L_3} - \left(\frac{\partial h_1}{\partial L_3}\right)_{L_1 L_2} \quad \text{and} \quad \frac{\partial h_i}{\partial L_2} = \left(\frac{\partial h_1}{\partial L_2}\right)_{L_1 L_3} - \left(\frac{\partial h_1}{\partial L_3}\right)_{L_1 L_2}$$

$[J]$ in this case becomes,

$$J = \begin{bmatrix} \sum \dfrac{\partial h_i}{\partial L_1} x_i & \sum \dfrac{\partial h_i}{\partial L_1} y_i \\[2mm] \sum \dfrac{\partial h_i}{\partial L_2} x_i & \sum \dfrac{\partial h_i}{\partial L_2} y_i \end{bmatrix} \tag{3.69}$$

$$\frac{\partial h_1}{\partial L_1} = \left(\frac{\partial h_1}{\partial L_1}\right)_{L_2 L_3} - \left(\frac{\partial h_1}{\partial L_3}\right)_{L_1 L_2} = 1$$

$$\frac{\partial h_2}{\partial L_1} = \left(\frac{\partial h_2}{\partial L_1}\right)_{L_2 L_3} - \left(\frac{\partial h_2}{\partial L_3}\right)_{L_1 L_2} = 0$$

$$\frac{\partial h_3}{\partial L_1} = \left(\frac{\partial h_3}{\partial L_1}\right)_{L_2 L_3} - \left(\frac{\partial h_3}{\partial L_3}\right)_{L_1 L_2} = -1$$

$$\frac{\partial h_1}{\partial L_2} = \left(\frac{\partial h_1}{\partial L_2}\right)_{L_1 L_3} - \left(\frac{\partial h_1}{\partial L_3}\right)_{L_1 L_2} = 0$$

$$\frac{\partial h_2}{\partial L_2} = \left(\frac{\partial h_2}{\partial L_2}\right)_{L_1 L_3} - \left(\frac{\partial h_2}{\partial L_3}\right)_{L_1 L_2} = 1$$

$$\frac{\partial h_3}{\partial L_2} = \left(\frac{\partial h_3}{\partial L_2}\right)_{L_1 L_3} - \left(\frac{\partial h_3}{\partial L_3}\right)_{L_1 L_2} = -1$$

$$\frac{\partial h_1}{\partial L_3} = \left(\frac{\partial h_1}{\partial L_3}\right)_{L_1 L_3} - \left(\frac{\partial h_1}{\partial L_3}\right)_{L_1 L_2} = 0$$

$$\frac{\partial h_2}{\partial L_3} = \left(\frac{\partial h_2}{\partial L_3}\right)_{L_1 L_2} - \left(\frac{\partial h_2}{\partial L_3}\right)_{L_1 L_2} = 0$$

$$\frac{\partial h_3}{\partial L_3} = \left(\frac{\partial h_3}{\partial L_3}\right)_{L_1 L_2} - \left(\frac{\partial h_3}{\partial L_3}\right)_{L_1 L_2} = 0$$

The weak formulation for a two-dimensional second order problem can now be recast by substituting the different terms and integrating using the numerical integration formulae for triangles.

Note that this method of derivation of the Jacobian and shape function derivatives using the area coordinates can be applied to length and volume coordinate derived shape functions as well.

3.7.3 Quadrilateral Isoparametric Element

The quadrilateral element can be represented with a bi-linear isoparametric element. The shape functions are same as for serendipity elements (see Eq. 3.55)

$$h_1(\xi, n) = \tfrac{1}{4}(1 - \xi)(1 - \eta)$$
$$h_2(\xi, n) = \tfrac{1}{4}(1 + \xi)(1 - \eta)$$
$$h_3(\xi, n) = \tfrac{1}{4}(1 + \xi)(1 + \eta)$$
$$h_4(\xi, n) = \tfrac{1}{4}(1 - \xi)(1 + \eta)$$

where $-1 \leq \xi \leq 1$ and $-1 \leq \eta \leq 1$

Thus,

$$\tilde{u} = h_1(\xi, \eta)u_1 + h_2(\xi, \eta)u_2 + h_3(\xi, n)u_3 + h_4(\xi, n)u_4$$

Similarly, the transformation equations apply

$$x = \sum_{i=1}^{4} h_i(\xi, \eta)x_i \quad \text{and} \quad y = \sum_{i=1}^{4} h_i(\xi, \eta)y_i$$

$$\frac{\partial h_i}{\partial \xi} = \frac{\partial h_i}{\partial x}\frac{\partial x}{\partial \xi} + \frac{\partial h_i}{\partial y}\frac{\partial y}{\partial \xi} \quad \text{and} \quad \frac{\partial h_i}{\partial \eta} = \frac{\partial h_i}{\partial x}\frac{\partial x}{\partial \eta} + \frac{\partial h_i}{\partial y}\frac{\partial y}{\partial \eta}$$

$$\begin{bmatrix} \dfrac{\partial h_i}{\partial \xi} \\ \dfrac{\partial h_i}{\partial \eta} \end{bmatrix} = \begin{bmatrix} \dfrac{\partial x}{\partial \xi} & \dfrac{\partial y}{\partial \xi} \\ \dfrac{\partial x}{\partial \eta} & \dfrac{\partial y}{\partial \eta} \end{bmatrix} \begin{bmatrix} \dfrac{\partial h_i}{\partial x} \\ \dfrac{\partial h_i}{\partial y} \end{bmatrix}$$

$$J = \begin{bmatrix} \dfrac{\partial x}{\partial \xi} & \dfrac{\partial y}{\partial \xi} \\ \dfrac{\partial x}{\partial \eta} & \dfrac{\partial y}{\partial \eta} \end{bmatrix} = \begin{bmatrix} J_{11} & J_{12} \\ J_{21} & J_{22} \end{bmatrix}$$

Following the same procedure in Sect. 3.7.2,

we have

$$
J = \begin{bmatrix}
\sum \dfrac{\partial h_i}{\partial \xi} x_i & \sum \dfrac{\partial h_i}{\partial \xi} y_i \\[2ex]
\sum \dfrac{\partial h_i}{\partial \eta} x_i & \sum \dfrac{\partial h_i}{\partial \eta} y_i
\end{bmatrix}
$$

Thus,

$$
\begin{aligned}
J_{11} &= -\tfrac{1}{4}(1-\eta)x_1 + \tfrac{1}{4}(1-\eta)x_2 + \tfrac{1}{4}(1+\eta)x_3 - \tfrac{1}{4}(1+\eta)x_4 \\
J_{12} &= -\tfrac{1}{4}(1-\eta)y_1 + \tfrac{1}{4}(1-\eta)y_2 + \tfrac{1}{4}(1+\eta)y_3 - \tfrac{1}{4}(1+\eta)y_4 \\
J_{21} &= -\tfrac{1}{4}(1-\xi)x_1 - \tfrac{1}{4}(1+\xi)x_2 + \tfrac{1}{4}(1+\xi)x_3 + \tfrac{1}{4}(1-\xi)x_4 \\
J_{22} &= -\tfrac{1}{4}(1-\xi)y_1 - \tfrac{1}{4}(1+\xi)y_2 + \tfrac{1}{4}(1+\xi)y_3 + \tfrac{1}{4}(1-\xi)y_4
\end{aligned}
$$

Also,

$$
\begin{bmatrix}
\dfrac{\partial h_i}{\partial x} \\[2ex]
\dfrac{\partial h_i}{\partial y}
\end{bmatrix}
=
\begin{bmatrix}
\dfrac{\partial x}{\partial \xi} & \dfrac{\partial y}{\partial \xi} \\[2ex]
\dfrac{\partial x}{\partial \eta} & \dfrac{\partial y}{\partial \eta}
\end{bmatrix}^{-1}
\begin{bmatrix}
\dfrac{\partial h_i}{\partial \xi} \\[2ex]
\dfrac{\partial h_i}{\partial \xi}
\end{bmatrix}
$$

where $\dfrac{\mathrm{d} h_i(\xi, \eta)}{\mathrm{d}x} = r_{11}\dfrac{\partial h_i}{\partial \xi} + r_{12}\dfrac{\partial h_i}{\partial \eta}$

Similarly,

$$
\frac{\mathrm{d} h_i(\xi, \eta)}{\mathrm{d}y} = r_{21}\frac{\partial h_i}{\partial \xi} + r_{22}\frac{\partial h_i}{\partial \eta},
$$

where $r_{ij},\ i = 1, 2, j = 1, 2$ are from the inverse matrix of

$$
[J] \quad \text{i.e.} \quad r = \begin{bmatrix} r_{11} & r_{12} \\ r_{21} & r_{22} \end{bmatrix} = [J]^{-1}
$$

The use of the Jacobian with numerical integration technique (the Gaussian quadrature) is normally employed. See applied numerical texts [5, 6] for detailed treatment of the Gauss–Legendre quadrature.

3.8 Exercises

1. Solve Eq. 3.6 using inverse matrix rule.
2. Derive the shape functions for four-node tetrahedral elements in Sect. 3.5.1 and eight-node brick elements as described in Sect. 3.5.2.
3. Equations 3.40 are called *Lagrange* shape functions because they can be obtained by products of one-dimensional shape functions. Obtain the sets of

one dimensional shape functions if the nodes are located at $x = -a$ and $x = a$; and $y = -b$ and $y = b$. *Hint*: Use the one-dimensional shape functions listed in Eqs. 3.12 and 3.13.

4. Using the relationship between the serendipity coordinate, r and the global coordinate, x, show that the shape function in Eq. 3.53 reduces to Eqs. 3.12 and 3.13.

5. Show that the shape functions in Eq. 3.55 will reduce to the Eq. 3.40 using the relationship between serendipity coordinates r,s and the global coordinates x, y.

6. Show that if the point m in Fig. 3.7 is located on the global coordinates, then the length coordinate shape functions (Eq. 3.57) are equal to the global coordinate shape functions of Eqs. 3.12 and 3.13.

7. Show that the area shape functions in Eq. 3.58 are equal to global shape functions of Eqs. 3.16–3.18.

8. Show that the derivatives of the linear triangular isoparametric shape functions in Sect. 3.6.4 are the same as for linear triangular shape functions in Sect. 3.4.1.

References

1. Surana KS (1987) Isoparametric finite elements. In: Kardestuncer H, Norrie D (eds) Finite element handbook. McGraw-Hill, New York
2. Stasa FL (1985) Applied finite element analysis for engineers. CBS Publishing Japan Ltd, New York
3. Reddy JN (2006) An introduction to the finite element method, 3rd edn. McGraw-Hill, New York
4. Cook RD (1987) Isoparametric families of elements. In: Kardestuncer H, Norrie D (eds) Finite element handbook. McGraw-Hill, New York
5. Carnahan B, Luther HA, Wilkes JO (1969) Applied numerical methods. Wiley, New York
6. Yang WY, Cao W, Chung TS, Morris J (2005) Applied numerical methods using MATLAB®. Wiley, New York

Chapter 4
Derivation of Element Matrices, Assembly and Solution of the Finite Element Equation

4.1 Derivation of Element Matrix for One-Dimensional Problems Using the Galerkin Method, Assembly and Solution

Example 1 Consider a one-dimensional second order ordinary differential equation

$$-\frac{\mathrm{d}}{\mathrm{d}x}\left(\frac{\mathrm{d}u}{\mathrm{d}x}\right) - f = 0; \quad 0 \le x \le 1; \quad u(0) = 0 \text{ and } u(1) = 0 \tag{4.1}$$

4.1.1 Weak Formulation

The weak formulation for an element will be

$$0 = \int_{x_i}^{x_i+1} w\left[-\frac{\mathrm{d}}{\mathrm{d}x}\left(\frac{\mathrm{d}u}{\mathrm{d}x}\right) - f\right]\mathrm{d}x \tag{4.2}$$

$$0 = \int_{x_i}^{x_i+1}\left[\frac{\mathrm{d}w}{\mathrm{d}x}\frac{\mathrm{d}u}{\mathrm{d}x} - wf\right]\mathrm{d}x - w\frac{\mathrm{d}u}{\mathrm{d}x}\Big|_{x_i}^{x_i+1} \tag{4.3}$$

where $w(\mathrm{d}u/\mathrm{d}x)|_{x_i}^{x_i+1}$ is the Neumann boundary condition.

Let us leave out the Neumann boundary condition term for now and work out the other terms in the equation, i.e.,

$$\int_{x_i}^{x_i+1}\left[\frac{\mathrm{d}w}{\mathrm{d}x}\frac{\mathrm{d}u}{\mathrm{d}x} - wf\right]\mathrm{d}x \tag{4.4}$$

O. Oluwole, *Finite Element Modeling for Materials Engineers Using MATLAB*®,
DOI: 10.1007/978-0-85729-661-0_4, © Springer-Verlag London Limited 2011

The approximation function is $\tilde{u} = h_1(x)u_1 + h_2(x)u_2$ (see Sect. 3.3.)
For the ith element

$$\tilde{u} = h_1(x)u_1 + h_2(x)u_{i+1} \tag{4.5}$$

The test (weighting function) is

$$w_1 = h_1(x) \quad \text{and} \quad w_2 = h_2(x) \text{ (Galerkin method)} \tag{4.6}$$

Substituting (4.5) and (4.6) in (4.4) yields

$$0 = \int_{x_i}^{x_{i+1}} \left[\left[\begin{Bmatrix} h_1' \\ h_2' \end{Bmatrix} (h_1' \quad h_2') \right] dx \begin{pmatrix} u_i \\ u_{i+1} \end{pmatrix} \right] - \int_{x_i}^{x_{i+1}} f \begin{pmatrix} h_1 \\ h_2 \end{pmatrix} dx \tag{4.7}$$

where h_1' and h_2' denote dh_1/dx and dh_2/dx, respectively.

From (3.15) and (3.16) $h_1^1 = -1/h_e$ and $h_2^1 = 1/h_e$.
Also,

$$h_i(x) = \frac{x_{i+1} - x}{h_e} \quad \text{and} \quad h_{i+1}(x) = \frac{x - x_i}{h_e}$$

wherever there is a product of the test function and the trial function, the resultant
is a matrix. The test function alone produces a vector. Thus, Eq. 4.7 becomes

$$0 = \frac{1}{h_e} \begin{pmatrix} 1 & -1 \\ -1 & 1 \end{pmatrix} \begin{pmatrix} u_i \\ u_{i+1} \end{pmatrix} - \frac{fh_e}{2} \begin{pmatrix} 1 \\ 1 \end{pmatrix} \tag{4.8}$$

This is the element finite element equation,

$$[K]^e[U]^e = [F]^e \tag{4.9}$$

It is now required to compute all the element equations.

4.1.2 Assembly of Element Equations

Let us divide the domain under consideration into three. Finite element equations
can be written in local (element) nodes as

Element 1

$$\begin{bmatrix} k_{11}^{(1)} & k_{12}^{(1)} \\ k_{21}^{(1)} & k_{22}^{(1)} \end{bmatrix} \begin{pmatrix} u_1^{(1)} \\ u_2^{(1)} \end{pmatrix} = \begin{bmatrix} F_1^{(1)} \\ F_2^{(1)} \end{bmatrix} \tag{4.10}$$

Element 2

$$\begin{bmatrix} k_{11}^{(2)} & k_{12}^{(2)} \\ k_{21}^{(2)} & k_{22}^{(2)} \end{bmatrix} \begin{pmatrix} u_1^{(2)} \\ u_2^{(2)} \end{pmatrix} = \begin{bmatrix} F_1^{(2)} \\ F_2^{(2)} \end{bmatrix} \tag{4.11}$$

Element 3

$$\begin{bmatrix} k_{11}^{(3)} & k_{12}^{(3)} \\ k_{21}^{(3)} & k_{22}^{(3)} \end{bmatrix} \begin{pmatrix} u_1^{(3)} \\ u_2^{(3)} \end{pmatrix} = \begin{bmatrix} F_1^{(3)} \\ F_2^{(3)} \end{bmatrix} \tag{4.12}$$

where $k^e ij$, u_i^e and F_i^e are the elements stiffness, variable and the force vectors, respectively.

To express the element equations in global nodes, the correspondent global nodes has to replace to element nodes using interelement continuity conditions stated below:

$$u_1^{(1)} = u_1$$

$$u_2^{(1)} = u_2$$

$$u_2^{(2)} = u_1^{(3)} = u_3$$

and

$$u_2^{(3)} = u_4$$

This correspondence between the element (local) nodes and the global nodes can be expressed in a Boolean connectivity matrix, B_{ij} = global node-number corresponding to the jth node of element i.

Thus,

$$[B_{ij}] = \begin{bmatrix} 1 & 2 \\ 2 & 3 \\ 3 & 4 \end{bmatrix} \tag{4.13}$$

The implication is that the coefficients of $k^{(e)}$ also add up (e.g.) (3,3) location of the assembled matrix has $k_{11}^{(3)}$ and k_{22}^{2}; (2,2) location has $k_{11}^{(2)}$ and $k_{22}^{(1)}$.

To express the element equations in the global nodes, each equation must be expanded such that the matrix and the vector size are the same as the total number of degrees of freedom in the system.

In the example under consideration, the total number of degrees of freedom is equal to 4. Expressing the element equations in global values give:

Element 1

$$\begin{bmatrix} k_{11}^{(1)} & k_{12}^{(1)} & 0 & 0 \\ k_{21}^{(1)} & k_{22}^{(1)} & 0 & 0 \\ 0 & 0 & 0 & 0 \\ 0 & 0 & 0 & 0 \end{bmatrix} \begin{bmatrix} u_1 \\ u_2 \\ u_3 \\ u_4 \end{bmatrix} = \begin{bmatrix} F_1^{(1)} \\ F_2^{(1)} \\ 0 \\ 0 \end{bmatrix} \tag{4.14}$$

Element 2

$$
\begin{bmatrix}
0 & 0 & 0 & 0 \\
0 & k_{11}^{(2)} & k_{12}^{(2)} & 0 \\
0 & k_{21}^{(2)} & k_{22}^{(12)} & 0 \\
0 & 0 & 0 & 0
\end{bmatrix}
\begin{bmatrix} u_1 \\ u_2 \\ u_3 \\ u_4 \end{bmatrix}
=
\begin{bmatrix} 0 \\ F_1^{(2)} \\ F_2^{(2)} \\ 0 \end{bmatrix}
\tag{4.15}
$$

Element 3

$$
\begin{bmatrix}
0 & 0 & 0 & 0 \\
0 & 0 & 0 & 0 \\
0 & 0 & k_{11}^{(3)} & k_{12}^{(3)} \\
0 & 0 & k_{21}^{(3)} & k_{22}^{(3)}
\end{bmatrix}
\begin{bmatrix} u_1 \\ u_2 \\ u_3 \\ u_4 \end{bmatrix}
=
\begin{bmatrix} 0 \\ 0 \\ F_1^{(3)} \\ F_2^{(3)} \end{bmatrix}
\tag{4.16}
$$

Assembling the element matrices into global matrix gives

$$
\begin{bmatrix}
k_1^{(1)} & k_{12}^{(1)} & 0 & 0 \\
k_{21}^{(1)} & k_{22}^{(1)} + k_{11}^{(2)} & k_{12}^{(2)} & 0 \\
0 & k_{21}^{(2)} & k_{22}^{(2)} + k_{12}^{(3)} & k_{12}^{(3)} \\
0 & 0 & k_{22}^{(3)} & k_{22}^{(3)}
\end{bmatrix}
\begin{bmatrix} U_1 \\ U_2 \\ U_3 \\ U_4 \end{bmatrix}
=
\begin{Bmatrix} F_1^{(1)} \\ F_2^{(1)} + F_1^{(2)} \\ F_2^{(2)} + F_1^{(3)} \\ F_2^{(3)} \end{Bmatrix}
\tag{4.17}
$$

At this point, the Neumann boundary condition is added to give:

$$
\begin{bmatrix}
k_1^{(1)} & k_{12}^{(1)} & 0 & 0 \\
k_{21}^{(1)} & k_{22}^{(1)} + k_{11}^{(2)} & k_{12}^{(2)} & 0 \\
0 & k_{21}^{(2)} & k_{22}^{(2)} + k_{12}^{(3)} & k_{12}^{(3)} \\
0 & 0 & k_{22}^{(3)} & k_{22}^{(3)}
\end{bmatrix}
\begin{bmatrix} U_1 \\ U_2 \\ U_3 \\ U_4 \end{bmatrix}
=
\begin{Bmatrix} F_1^{(1)} - u'(0) \\ F_2^{(1)} + F_1^{(2)} \\ F_2^{(2)} + F_1^{(3)} \\ F_2^{(3)} + u'(1) \end{Bmatrix}
\tag{4.18}
$$

This is the generalized global finite element equation for the second order ordinary differential equation Eq. 4.1 given above, k_{ij}^e and F_i^e will have to be determined.

4.1.3 Imposition of Boundary Conditions

In solving the finite element equation it is important to solve for the nodal values (primary variable) first. This is done by imposing the Dirichlet (essential) boundary conditions first in the solving of this finite element equation. In this example, u(0) = 0 and u(1) = 0 and the domain $0 \leq x \leq 1$. Substituting them into the global finite element equation gives

$$
\begin{bmatrix}
1 & 0 & 0 & 0 \\
K_{21}^{(1)} & K_{22}^{(1)} + K_{11}^{(2)} & K_{12}^{(2)} & 0 \\
0 & K_{21}^{(2)} & K_{22}^{(2)} + K_{12}^{(3)} & K_{12}^{(3)} \\
0 & 0 & 0 & 1
\end{bmatrix}
\begin{bmatrix} u_1 \\ u_2 \\ u_3 \\ u_4 \end{bmatrix}
=
\begin{bmatrix} 0 \\ F_1^{(2)} + F_1^{(2)} \\ F_2^{(2)} + F_1^{(3)} \\ 0 \end{bmatrix}
\tag{4.19}
$$

Since all the values in this resultant matrix are known all the nodal values (u_2 and u_3) can be obtained by solving the resultant simultaneous equation.

4.1.4 Obtaining Neumann Boundary Conditions at X = 0 and X = l

In the example under consideration, the Neumann conditions $u'(0)$ and $u'(1)$ are not provided. These can be obtained by substituting back into Eq. 4.18 and solving. $u'(0)$ and $u'(1)$ can be obtained from the resultant set of simultaneous equations.

4.2 Derivation of Element Matrix for Two-Dimensional Problems Using the Galerkin Method

Let us use the poisson equation in two-dimension expressed as:

$$-\left[\frac{\partial}{\partial x}\left(\frac{\partial u}{\partial x}\right)+\frac{\partial}{\partial y}\left(\frac{\partial u}{\partial y}\right)\right] = Q \tag{4.20}$$

The weak formulation is

$$0 = \int_\Omega \left[\frac{\partial w}{\partial x}\frac{\partial u}{\partial x}+\frac{\partial w}{\partial y}\frac{\partial u}{\partial y}\right]\mathrm{d}x\mathrm{d}y = \int_\Omega wQ\mathrm{d}x\mathrm{d}y + \oint_\Gamma w\left(\frac{\partial u}{\partial x}n_x + \frac{\partial u}{\partial y}n_y\right)\mathrm{d}s \tag{4.21}$$

Equation 4.24 can be rewritten as

$$[K][U] = [F],$$

where

$$[K]^e = \int_\Omega \left(\frac{\partial w}{\partial x}\frac{\partial u}{\partial x}+\frac{\partial w}{\partial y}\frac{\partial u}{\partial y}\right)\mathrm{d}\Omega \tag{4.22}$$

$$[F]^e = \oint_{\Omega^e} WQ\mathrm{d}\Omega + \oint w\frac{\partial u}{\partial n}\mathrm{d}s \tag{4.23}$$

4.2.1 Using Triangular Discretization

Substituting (3.33), (3.34), (3.35) and (3.36) into $[K]^e$ gives

$$[K]^e = \int_{\Omega^e} \left[\frac{1}{2A} \begin{pmatrix} y_2 - y_3 \\ y_3 - y_1 \\ y_1 - y_2 \end{pmatrix} \left[\frac{1}{2A} (y_2 - y_3 \quad y_3 - y_1 \quad y_1 - y_2) \right] \begin{pmatrix} u_1 \\ u_2 \\ u_3 \end{pmatrix} \right.$$

$$\left. + \frac{1}{4A^2} \begin{pmatrix} x_3 - x_2 \\ x_1 - x_3 \\ x_2 - x_1 \end{pmatrix} (x_3 - x_2 \quad x_1 - x_3 \quad x_2 - x_1) \begin{pmatrix} u_1 \\ u_2 \\ u_3 \end{pmatrix} \right] d\Omega \qquad (4.24)$$

$$= \frac{1}{4A} \begin{bmatrix} (y_2 - y_3)^2 & (y_2 - y_3)(y_3 - y_1) & (y_2 - y_3)(y_1 - y_2) \\ (y_3 - y_1)(y_2 - y_3) & (y_3 - y_1)^2 & (y_3 - y_1)(y_1 - y_2) \\ (y_1 - y_2)(y_2 - y_3) & (y_3 - y_1)(y_1 - y_2) & (y_1 - y_2)^2 \end{bmatrix} \begin{pmatrix} u_1 \\ u_2 \\ u_3 \end{pmatrix}$$

$$+ \frac{1}{4A} \begin{bmatrix} (x_3 - x_2)^2 & (x_3 - x_2)(x_1 - x_3) & (x_3 - x_2)(x_2 - x_1) \\ (x_3 - x_2)(x_1 - x_3) & (x_1 - x_3)^2 & (x_1 - x_3)(x_2 - x_1) \\ (x_2 - x_1)(x_3 - x_2) & (x_2 - x_1)(x_1 - x_3) & (x_2 - x_1)^2 \end{bmatrix} \begin{pmatrix} u_1 \\ u_2 \\ u_3 \end{pmatrix}$$

$$(4.25)$$

Thus

$$[K]^e = \begin{bmatrix} k_{11} & k_{12} & k_{33} \\ k_{21} & k_{22} & k_{23} \\ k_{31} & k_{32} & k_{33} \end{bmatrix} \qquad (4.26)$$

where

$$k_{11} = \frac{1}{4A} \left[(x_3 - x_2)^2 + (y_2 - y_3)^2 \right]$$

$$k_{12} = \frac{1}{4A} [(x_3 - x_2)(x_1 - x_3) + (y_2 - y_3)(y_3 - y_1)]$$

$$k_{13} = \frac{1}{4A} [(x_3 - x_2)(x_2 - x_1) + (y_2 - y_3)(y_1 - y_2)]$$

$$k_{21} = k_{12}$$

$$k_{22} = \frac{1}{4A} \left[(x_1 - x_3)^2 + (y_3 - y_1)^2 \right]$$

$$k_{22} = \frac{1}{4A} [(x_1 - x_3)(x_2 - x_1) + (y_3 - y_1)(y_1 - y_2)]$$

$$k_{23} = \frac{1}{4A} [(x_1 - x_3)(x_2 - x_1) + (y_3 - y_1)(y_1 - y_2)]$$

$$k_{31} = k_{13}$$

$$k_{32} = k_{23}$$

$$k_{33} = \frac{1}{4A}\left[(x_2 - x_1)^2 + (y_1 - y_2)^2\right] \tag{4.27}$$

$$[F]^e = \int_{\Omega^e} \left\{\begin{array}{c} h_1 \\ h_2 \\ h_3 \end{array}\right\} Q\,d\Omega + \oint_{\Gamma^e} w\frac{du}{dn}ds \tag{4.28}$$

4.2.2 Using Bilinear Elements

Using bilinear rectangular elements for the same problem of Eq. 4.23 and substituting bilinear rectangular equations of (3.42), (3.43) and (3.44) into the equation gives:

$$[K]^e = \int_{\Omega^e} \left(k_x \left\{\begin{array}{c} \frac{dh_1}{dx} \\ \frac{dh_2}{dx} \\ \frac{dh_3}{dx} \\ \frac{dh_4}{dx} \end{array}\right\} \left[\begin{array}{cccc} \frac{dh_1}{dx} & \frac{dh_2}{dx} & \frac{dh_3}{dx} & \frac{dh_4}{dx} \end{array}\right) \left(\begin{array}{c} u_1 \\ u_2 \\ u_3 \\ u_4 \end{array}\right)\right.$$

$$\left. + k_y \left\{\begin{array}{c} \frac{dh_1}{dy} \\ \frac{dh_2}{dy} \\ \frac{dh_3}{dy} \\ \frac{dh_4}{dy} \end{array}\right\} \left[\begin{array}{cccc} \frac{dh_1}{dy} & \frac{dh_2}{dy} & \frac{dh_3}{dy} & \frac{dh_4}{dy} \end{array}\right] \right) d\Omega \tag{4.29}$$

Matrix will be 4×4, thus:

$$[K]^e = \begin{bmatrix} k_{11} & k_{12} & k_{13} & k_{14} \\ k_{21} & k_{22} & k_{23} & k_{24} \\ k_{31} & k_{32} & k_{33} & k_{34} \\ k_{41} & k_{42} & k_{43} & k_{44} \end{bmatrix} \tag{4.30}$$

Also, for the bilinear rectangular element,

$$\begin{aligned} K_{11} &= k_{22} = K_{33} = k_{44} \\ K_{23} &= k_{14} \\ K_{24} &= k_{13} \\ K_{34} &= k_{12} \end{aligned} \tag{4.31}$$

Thus

$$[K^e] = \begin{bmatrix} k_{11} & k_{12} & k_{13} & k_{14} \\ k_{12} & k_{11} & k_{14} & k_{13} \\ k_{13} & k_{14} & k_{11} & k_{12} \\ k_{14} & k_{13} & k_{12} & k_{11} \end{bmatrix} \qquad (4.32)$$

$$\begin{aligned} k_{11} &= \frac{a^2 + b^2}{3ab} \\ k_{12} &= \frac{a^2 - 2b^2}{6ab} \\ k &= -\frac{a^2 + b^2}{6ab} \\ k_{14} &= \frac{a^2 - 2b^2}{6ab} \end{aligned} \qquad (4.33)$$

$$[F]^e = \int\limits_{-a}^{a} \int\limits_{-b}^{b} \begin{Bmatrix} h_1 \\ h_2 \\ h_3 \\ h_4 \end{Bmatrix} Q + \oint\limits_{\Gamma^e} w \frac{du}{dn} ds \qquad (4.34)$$

$$\oint\limits_{\Gamma^e} w \frac{du}{dn} ds = \oint\limits_{\Gamma^e} w \left[\frac{du}{dx} n_x + \frac{du}{dy} n_y \right] dxdy \qquad (4.35)$$

4.3 Derivation of Element Matrix for Three-Dimensional Problems Using the Galerkin Method

Let us consider the three-dimensional Poisson Equation

$$\nabla^2 u = Q \qquad (4.36)$$

The weak formulation will be

$$-\int\limits_{\Omega} \left(\frac{\partial w}{\partial x} \frac{\partial u}{\partial x} + \frac{\partial w}{\partial y} \frac{\partial u}{\partial y} + \frac{\partial w}{\partial z} \frac{\partial u}{\partial z} \right) d\Omega - \int\limits_{\Omega} wQ\, d\Omega + \oint\limits_{\Gamma} w \frac{\partial u}{\partial n} ds$$

Thus using a four node tetrahedral element,

$$[k^e] = \int_{\Omega^e} \left(\left(\begin{Bmatrix} \dfrac{dh_1}{dx} \\[4pt] \dfrac{dh_2}{dx} \\[4pt] \dfrac{dh_3}{dx} \\[4pt] \dfrac{dh_4}{dx} \end{Bmatrix} \begin{bmatrix} \dfrac{dh_1}{dx} & \dfrac{dh_2}{dx} & \dfrac{dh_3}{dx} & \dfrac{dh_4}{dx} \end{bmatrix} + \begin{bmatrix} \dfrac{dh_1}{dy} \\[4pt] \dfrac{dh_2}{dy} \\[4pt] \dfrac{dh_3}{dy} \\[4pt] \dfrac{dh_4}{dy} \end{bmatrix} \begin{bmatrix} \dfrac{dh_1}{dy} & \dfrac{dh_2}{dy} & \dfrac{dh_3}{dy} & \dfrac{dh_4}{dy} \end{bmatrix} \right. \right.$$

$$\left. \left. + \begin{Bmatrix} \dfrac{dh_1}{dz} \\[4pt] \dfrac{dh_2}{dz} \\[4pt] \dfrac{dh_3}{dz} \\[4pt] \dfrac{dh_4}{dz} \end{Bmatrix} \begin{bmatrix} \dfrac{dh_1}{dz} & \dfrac{dh_2}{dz} & \dfrac{dh_3}{dz} & \dfrac{dh_4}{dz} \end{bmatrix} \right) d\Omega \right) \tag{4.37}$$

$$[F]^e = \int_{\Omega} wQ d\Omega + \oint_{\Gamma} w \frac{\partial u}{\partial n} ds.$$

The same workings as done for the two-dimensional problem (see Sect. 4.2) would be done here to get the element equations.

$$\int_{\Omega} wQ\, d\Omega = \int_{\Omega} \begin{Bmatrix} h_1 \\ h_2 \\ h_3 \\ h_4 \end{Bmatrix} Q\, d\Omega$$

$\oint_{\Gamma} w(du/dn)ds$ can be treated looking at boundary conditions on the four sides of the tetrahedral, placing the boundary conditions as specified for the nodes.

4.4 Transient Problems

Let us take for an example one dimensional transient heat conduction in a domain Ω. The equation is

$$\frac{\partial u}{\partial t} = \frac{1}{\alpha} \left(\frac{\partial^2 u}{\partial x^2} \right) \quad \text{in } \Omega, 0 < x < 1 \tag{4.38}$$

where α is equal to $k/(\rho C_p)$, k is thermal conductivity and C_p is the specific heat capacity at constant pressure. Heat generation (Heat source, Q) is not included here.

The weak formulation is:

$$\int_{\Omega} w\frac{du}{dt}dx = -\frac{1}{\alpha}\int_{\Omega}\left(\frac{dw}{dx}\frac{du}{dx}\right)dx + \frac{1}{\alpha}\oint_{\Gamma_n} w\frac{\partial u}{\partial x}d\Gamma \qquad (4.39)$$

The element equation can be expressed as $[L^e][\dot{u}]^t + [K^e][u]^t = [F^e]^t$ which is different from the steady-state form in that the solution is time dependent. However, the matrices $[L^e]$ and $[K^e]$ are not time dependent.

Using the shape functions for triangular elements derived in Chap. 3 for the element equation formulation for $[L^e]$

$$[L^e] = \int \begin{Bmatrix} h_1 \\ h_2 \\ h_3 \end{Bmatrix} (\begin{array}{ccc} h_1 & h_2 & h_3 \end{array})dx \begin{pmatrix} \dot{u}_1 \\ \dot{u}_2 \\ \dot{u}_3 \end{pmatrix} \qquad (4.40)$$

where $\dot{u} = $ variation of u with time

$$[L^e] = \frac{h_e}{6}\begin{bmatrix} 2 & 1 \\ 1 & 2 \end{bmatrix} \qquad (4.41a)$$

Note For *two-dimension* problems, $[L^e]$ will be

$$\frac{A}{12}\begin{bmatrix} 2 & 1 & 1 \\ 1 & 2 & 1 \\ 1 & 1 & 2 \end{bmatrix} \quad \text{for linear triangular meshing} \qquad (4.41b)$$

and

$$\frac{A}{36}\begin{bmatrix} 4 & 2 & 1 & 2 \\ 2 & 4 & 2 & 1 \\ 1 & 2 & 4 & 2 \\ 2 & 1 & 2 & 4 \end{bmatrix} \quad \text{for rectangular element meshing} \qquad (4.41c)$$

4.4.1 Time Integration Method for Transient Problems

Finite difference methods are used for the time derivatives. Commonly used methods are; the forward difference, backward difference and the Crank–Nicolson method. Detailed analyses of these methods are obtained in standard Numerical Methods texts [1, 2] and will not be discussed here but the backward difference method will be applied in the examples. The backward difference and Crank–Nicolson methods are unconditionally stable.

4.4.1.1 Backward Difference Method

In this method the finite element and the time derivative can be written as:

$$[L]\,\{\dot{u}\}^{t+\Delta t} = [K]\,\{u\}^{t+\Delta t} + [F]^{t+\Delta t} \tag{4.42}$$

with

$$\{\dot{U}\}^{t+\Delta t} = \frac{\{U\}^{t+\Delta t} - \{U\}^{t}}{\Delta t}, \tag{4.43}$$

respectively. Substituting (4.43) in (4.42) gives the finite element equation;

$$([L] + \Delta t[K])\{U\}^{t+\Delta t} = \Delta t[F]^{t+\Delta t} + [L]\{U\}^{t} \tag{4.44}$$

Solving the one-dimensional problem of (Sect. 4.4) using four element mesh, the backward difference method and taking $\alpha = 1$, the assembled equations follow the procedures used for Eq. 4.1. Thus:

$$\text{The global } [L] = \frac{2}{3}\begin{bmatrix} 2 & 1 & 0 & 0 & 0 \\ 1 & 4 & 1 & 0 & 0 \\ 0 & 1 & 4 & 1 & 0 \\ 0 & 0 & 1 & 4 & 1 \\ 0 & 0 & 0 & 1 & 2 \end{bmatrix} \tag{4.45}$$

$$[K] = \frac{1}{0.25}\begin{bmatrix} 1 & -1 & 0 & 0 & 0 \\ -1 & 2 & -1 & 0 & 0 \\ 0 & -1 & 2 & -1 & 0 \\ 0 & 0 & -1 & 2 & -1 \\ 0 & 0 & 0 & -1 & 1 \end{bmatrix} \tag{4.46}$$

taking $U_1 = 400$ and $U_4 = 25$, i.e., one end heated up and the other kept at room temperature.

$$\{F\} = \begin{Bmatrix} F_1 \\ 0 \\ 0 \\ 0 \\ F_5 \end{Bmatrix} \quad \text{also } \{U\}^{t=0} = 25 \tag{4.47}$$

Substituting (4.45) through (4.47) in

$$([L] + \Delta t[K])[U]^{t+\Delta t} = \Delta t\{F\}^{t+\Delta t} + [L]\{U\}^{t}$$

gives

$$= \begin{bmatrix} \frac{4}{3}+4\Delta t & \frac{2}{3}-4\Delta t & 0 & 0 & 0 \\ \frac{2}{3}-4\Delta t & \frac{8}{3}+8\Delta t & \frac{2}{3}-4\Delta t & 0 & 0 \\ 0 & \frac{2}{3}-4\Delta t & \frac{8}{3}+8\Delta t & \frac{2}{3}-4\Delta t & 0 \\ 0 & 0 & \frac{2}{3}-4\Delta t & \frac{8}{3}+8\Delta t & \frac{2}{3}-4\Delta t \\ 0 & 0 & 0 & \frac{2}{3}-4\Delta t & \frac{4}{3}+4\Delta t \end{bmatrix} \begin{Bmatrix} U_1^{t+\Delta t} \\ U_2^{t+\Delta t} \\ U_3^{t+\Delta t} \\ U_4^{t+\Delta t} \\ U_5^{t+\Delta t} \end{Bmatrix}$$

$$\qquad\qquad\qquad\qquad\qquad\qquad\qquad\qquad\qquad\qquad\qquad\qquad\qquad (4.48a)$$

$$= \begin{Bmatrix} \frac{4}{3}U_1^t + \frac{2}{3}U_2^t + \frac{1}{3}F_1\Delta t \\ \frac{2}{3}U_1^t + \frac{8}{3}U_2^t + \frac{2}{3}U_3^t \\ \frac{2}{3}U_2^t + \frac{8}{3}U_3^t + \frac{2}{3}U_4^t \\ \frac{2}{3}U_3^t + \frac{8}{3}U_4^t + \frac{2}{3}U_5^t \\ \frac{2}{3}U_4^t + \frac{4}{3}U_5^t + F_5\Delta t \end{Bmatrix}$$

Applying time step size of $\Delta t = 1/3$ and using the boundary conditions give

$$\begin{bmatrix} 1 & 0 & 0 & 0 & 0 \\ -\frac{2}{3} & \frac{16}{3} & -\frac{2}{3} & 0 & 0 \\ 0 & -\frac{2}{3} & \frac{16}{3} & -\frac{2}{3} & 0 \\ 0 & 0 & -\frac{2}{3} & \frac{16}{3} & -\frac{2}{3} \\ 0 & 0 & 0 & 0 & 1 \end{bmatrix} \begin{Bmatrix} U_1^{t+\Delta t} \\ U_2^{t+\Delta t} \\ U_3^{t+\Delta t} \\ U_4^{t+\Delta t} \\ U_5^{t+\Delta t} \end{Bmatrix} = \begin{Bmatrix} 400 \\ \frac{2}{3}U_1^t + \frac{8}{3}U_2^t + \frac{2}{3}U_3^t \\ \frac{2}{3}U_2^t + \frac{8}{3}U_3^t + \frac{2}{3}U_4^t \\ \frac{2}{3}U_3^t + \frac{8}{3}U_4^t + \frac{2}{3}U_5^t \\ 25 \end{Bmatrix} \qquad (4.48b)$$

Substituting $t = 0$ in Eqs. 4.48a, 4.48b gives

$$\left\{ U_1^{1/3} \quad U_2^{1/3} \quad U_3^{1/3} \quad U_4^{1/3} \right\} = \{ 400 \quad 58 \quad 7 \quad 4 \quad 25 \} \qquad (4.49)$$

$U_i^{1/3}$ = nodal temperature at time $t = 1/3$.

The iteration can be continued until the desired time solution is obtained by re-substituting $t = 1/3$ into Eq. 4.48b.

4.5 Derivation of Matrix Equations for Axisymmetric Problems

We will take an example of the Poisson equation in cylindrical (polar) coordinate system. The equations can be written as follows:

1. $$\left(\frac{\partial^2 u}{\partial r^2} + \frac{1}{r}\frac{\partial u}{\partial r}\right) = f \quad \text{in one-dimension} \tag{4.50}$$

2. $$\frac{\partial^2 u}{\partial r^2} + \frac{1}{r}\frac{\partial u}{\partial r} + \frac{1}{r^2}\frac{\partial^2 u}{\partial \theta^2} = f \quad \text{in two-dimensions} \tag{4.51}$$

3. $$\frac{\partial^2 u}{\partial r^2} + \frac{1}{r}\frac{\partial u}{\partial r} + \frac{1}{r^2}\frac{\partial^2 u}{\partial \theta^2} + \frac{\partial^2 u}{\partial z^2} = f \quad \text{in three-dimensions} \tag{4.52}$$

r, θ and z are the radial, circumferential and axial axes, respectively.

In axisymmetric problems, u is independent of θ and as such the three-dimensional equation reduces to

$$\frac{d^2 u}{dr^2} + \frac{1}{r}\frac{du}{dr} + \frac{d^2 u}{dz^2} = f \tag{4.53}$$

The weak formulation can be obtained as follows:

$$\int_\Omega \left\{ w\left(\frac{d^2 u}{dr^2} + \frac{1}{r}\frac{du}{dr} + \frac{d^2 u}{dz^2}\right) - (wf) \right\} d\Omega = 0 \tag{4.54}$$

Collecting the values in r together gives a typical heat conduction problem with heat generation.

$$\int_\Omega \left\{ w\left(\frac{1}{r}\frac{d}{dr}\left(r\frac{du}{dr}\right) + \frac{d^2 u}{dz^2}\right) - (wf) \right\} d\Omega = 0 \tag{4.55}$$

Using the relationship

$$\int_r \int_\theta \int_z B(r,z)\,dr\,d\theta\,dz = 2\pi \int_r \int_z rB(r,z)\,dr\,dz \tag{4.56}$$

the equation becomes

$$2\pi \int_r \int_z w\left\{ \frac{d}{dr}\left(r\frac{du}{dr}\right) - f + r\frac{d^2 u}{dz^2} \right\} d\Omega$$

$$= 2\pi \int_r \int_z \left(-r\frac{dw}{dr}\frac{du}{dr} - wf - r\frac{dw}{dz}\frac{du}{dz} \right) dr\,dz + \oint_\Gamma rw\frac{du}{dn}\,d\Gamma \tag{4.57}$$

Since Eq. 4.57 has become a two-dimensional problem, the two-dimensional shape functions will be applied in the derivation of the finite element matrix equation. Using linear triangular elements, the equation becomes:

$$[K^e] = 2\pi \int_r \int_z r \left(\begin{Bmatrix} \dfrac{dh_1}{dr} \\[4pt] \dfrac{dh_2}{dr} \\[4pt] \dfrac{dh_3}{dr} \end{Bmatrix} \begin{Bmatrix} \dfrac{dh_1}{dr} & \dfrac{dh_2}{dr} & \dfrac{dh_3}{dr} \end{Bmatrix} \right.$$

$$\left. + \begin{Bmatrix} \dfrac{dh_1}{dz} \\[4pt] \dfrac{dh_2}{dz} \\[4pt] \dfrac{dh_3}{dz} \end{Bmatrix} \left(\dfrac{dh_1}{dz} \quad \dfrac{dh_2}{dz} \quad \dfrac{dh_3}{dz} \right) \right) dr\, dz \tag{4.58}$$

also

$$\int_r \int_z r\, dr\, dz = A r_c \tag{4.59}$$

A is the area of the element and r_c is the r coordinate value of the centroid of the triangular element.

Thus,

$$[K^e] = 2\pi r_c \begin{bmatrix} K_{11} & K_{12} & K_{13} \\ K_{21} & K_{22} & K_{23} \\ K_{31} & K_{32} & K_{33} \end{bmatrix} \tag{4.60}$$

where

$$K_{11} = \frac{1}{4A}\left[(r_3 - r_2)^2 + (z_2 - z_3)^2\right]$$

$$K_{13} = \frac{1}{4A}[(r_3 - r_2)(r_2 - r_1) + (z_2 - z_3)(z_1 - z_2)]$$

$$K_{12} = \frac{1}{4A}[(r_3 - r_2)(r_1 - r_3) + (z_2 - z_3)(z_3 - z_1)]$$

$$K_{22} = \frac{1}{4A}\left[(r_1 - r_3)^2 + (z_3 - z_1)^2\right]$$

$$K_{23} = \frac{1}{4A}[(r_1 - r_3)(r_2 - r_1) + (z_3 - z_1)(z_1 - z_2)]$$

$$K_{33} = \frac{1}{4A}\left[(r_2 - r_1)^2 + (z_1 - z_2)^2\right]$$

$$K_{12} = K_{21}$$

$$K_{13} = K_{31}$$

$$K_{23} = K_{32}$$

Fig. 4.1 A triangular plate

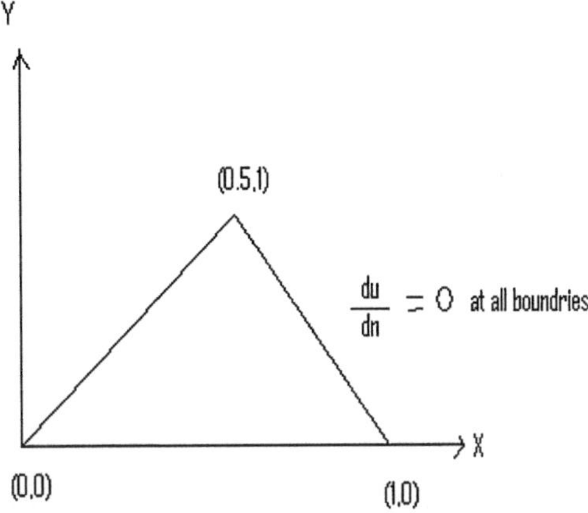

The flux $[F^e]-$d can be solved like the two-dimensional problem. The same procedure can be done for any other physical problem e.g. axysymmetric stress analysis by Obtaining the element equations from the governing equations.

4.6 Sample Solutions on Elements Matrix Computation, Assembly and Solution

Problem 4.1 Let us compute the element matrix for Poisson's equation for the triangular plate shown in Fig. 4.1. Flux at all boundaries $= 0$ and $a = 1$.

Solution Let us take the plate as a single triangular element.

$$[K^e] = \begin{bmatrix} k_{11} & k_{12} & k_{13} \\ k_{21} & k_{22} & k_{23} \\ k_{31} & k_{32} & k_{33} \end{bmatrix}$$

$$A = \frac{1}{2}\det \begin{vmatrix} 1 & x_1 & y_1 \\ 1 & x_2 & y_2 \\ 1 & x_3 & y_3 \end{vmatrix}$$

$$= \frac{1}{2}[1(x_2y_3 - x_3y_2) + 1(x_3y_1 - x_1y_3) + (x_1y_2 - x_2y_1)]$$

$$= \frac{1}{2}[(1 - (0.5)(0) + 1(0.5)(0) - 0(0) + (0) - (0)]$$

$$A = \frac{1}{2}$$

Fig. 4.2 A rectangular domain with four triangular meshes

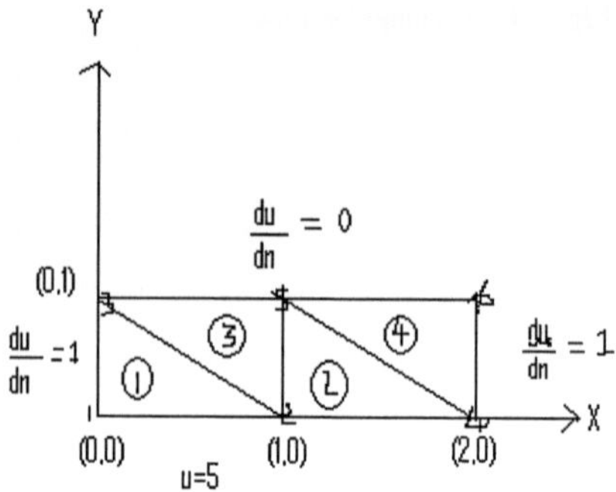

The values of the $[K^e]$ matrix can be calculated with the equations derived in this chapter for two-dimensional problems.

Thus:

$$K_{11} = \frac{1}{4A}\left[(0.5 - 1)^2 + (0 - 1)^2\right] = (-0.5)^2 + 1^2 = \left(\frac{1}{4}\right) = 0.625$$

$$K_{12} = [(-0.5)(0.5) + (0 - 1)(1 - 0)] = \frac{1}{2}(-0.25 - 1) = -0.625$$

$$K_{13} = \frac{1}{2}[(-0.5)(1) + (-1)(0)] = \frac{1}{2}(-0.5) = -0.25$$

$$K_{21} = K_{12}$$

$$K_{22} = \frac{1}{2}\left[(-0.5)^2 + (1)^2\right] = +\frac{1.25}{2} = 0.625$$

$$K_{23} = \frac{1}{2}[(-0.5)(1) + (1)(0)] = \frac{1}{2}(-0.5) = -0.5$$

$$K_{31} = K_{13}$$

$$K_{32} = K_{23}$$

$$K_{33} = \frac{1}{2}\left[(1)^2(0)\right] = 0.5$$

Thus,

$$[K^e] = \begin{bmatrix} 0.625 & -0.625 & -0.25 \\ -0.625 & 0.625 & -0.25 \\ -0.25 & -0.25 & 0.5 \end{bmatrix}$$

Problem 4.2 Consider a rectangular domain discretized into four linear triangular elements as shown. The Laplacian equation works within this domain. Find nodal temperatures if there is no heat generation, i.e., $Q = 0$ (Fig. 4.2).

Solution In Problem 4.1, we successfully got an element stiffness matrix $[K^e]$. In this problem (4.2), we need to find each element matrix first. To do this we must first sort out the local and global node numbering. We can number the elements as shown below. The local numbering helps in calculating the $[K^e]$ matrices.

As mentioned, the local nodes help in calculating the $[K^e]$ matrices, i.e., assigning of shape functions. The global nodes are useful in identifying the element position in the overall domain and global stiffness matrix.

Using Eq. 4.26;

$$[K^e] = \begin{bmatrix} 1 & -0.5 & -0.5 \\ -0.5 & 0.5 & 0 \\ -0.5 & 0 & 0.5 \end{bmatrix}$$

The element matrices are the same in this situation.

Re-structuring the element matrices in global nodes gives;

$$[K^1] = \begin{bmatrix} 1 & -0.5 & -0.5 & 0 & 0 & 0 \\ -0.5 & 0.5 & 0 & 0 & 0 & 0 \\ -0.5 & 0 & 0.5 & 0 & 0 & 0 \\ 0 & 0 & 0 & 0 & 0 & 0 \\ 0 & 0 & 0 & 0 & 0 & 0 \\ 0 & 0 & 0 & 0 & 0 & 0 \end{bmatrix}$$

Notice the placement of values at the local nodes (1, 2, 3) in the global nodes (1, 2, 3), respectively.

$$[K^2] = \begin{bmatrix} 0 & 0 & 0 & 0 & 0 & 0 \\ 0 & 1 & 0 & -0.5 & -0.5 & 0 \\ 0 & 0 & 0 & 0 & 0 & 0 \\ 0 & -0.5 & 0 & 0.5 & 0 & 0 \\ 0 & -0.5 & 0 & 0 & 0.5 & 0 \\ 0 & 0 & 0 & 0 & 0 & 0 \end{bmatrix}$$

Notice the placement of values at the local nodes (1, 2, 3) in the global nodes (2, 4, 5), respectively.

$$[K^3] = \begin{bmatrix} 0 & 0 & 0 & 0 & 0 & 0 \\ 0 & 0.5 & 0 & 0 & -0.5 & 0 \\ 0 & 0 & 0.5 & 0 & -0.5 & 0 \\ 0 & 0 & 0 & 0 & 0 & 0 \\ 0 & -0.5 & -0.5 & 0 & 1 & 0 \\ 0 & 0 & 0 & 0 & 0 & 0 \end{bmatrix}$$

Notice the placement of values at the local nodes (1, 2, 3) in the global nodes (5, 3, 2), respectively

$$[K^4] = \begin{bmatrix} 0 & 0 & 0 & 0 & 0 & 0 \\ 0 & 0 & 0 & 0 & 0 & 0 \\ 0 & 0 & 0 & 0 & 0 & 0 \\ 0 & 0 & 0 & 0.5 & 0 & -0.5 \\ 0 & 0 & 0 & 0 & 0.5 & -0.5 \\ 0 & 0 & 0 & -0.5 & -0.5 & 1 \end{bmatrix}$$

Notice the placement of values at the local nodes (1, 2, 3) in the global nodes (6, 5, 4), respectively.

Assembling all the elemental matrices into the global matrix gives.

$$[K] = \begin{bmatrix} 1 & -0.5 & -0.5 & 0 & 0 & 0 \\ -0.5 & 2 & 0 & -0.5 & -1.0 & 0 \\ -0.5 & 0 & 1 & 0 & -0.5 & 0 \\ 0 & -0.5 & 0 & 1 & 0 & -0.5 \\ 0 & -1.0 & -0.5 & 0 & 2 & -0.5 \\ 0 & 0 & 0 & -0.5 & -0.5 & 1 \end{bmatrix}$$

In this particular example, the global matrix is occupied by these local values:

$$K_{11} = K^1_{11}$$

$$K_{12} = K^1_{12}$$

$$K_{13} = K^1_{13}$$

$$K_{21} = K^1_{21}$$

$$K_{22} = K^1_{22} + K^2_{11} + K^3_{33}$$

$$K_{23} = K^1_{23} + K^3_{32}$$

$$K_{24} = K^2_{12}$$

$$K_{25} = K^2_{13} + K^3_{31}$$

$$K_{26} = 0$$

$$K_{31} = K^1_{31}$$

$$K_{32} = K^1_{32} + K^3_{23}$$

$$K_{33} = K^1_{33} + K^3_{22}$$

$$K_{34} = 0$$

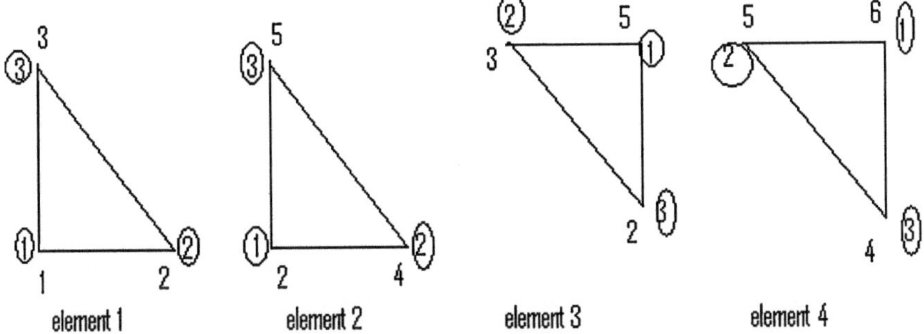

Fig. 4.3 Local and global numbering of Fig. 4.2. Ⓘ represents local node numbers; $i =$ global node numbers

$$K_{35} = K_{21}^3$$

$$K_{36} = 0$$

$$K_{41} = 0$$

$$K_{42} = K_{21}^2$$

$$K_{43} = 0$$

$$K_{44} = K_{22}^2 + K_{33}^4$$

$$K_{45} = K_{23}^2 + K_{32}^4$$

$$K_{46} = K_{31}^4$$

$$K_{51} = 0$$

$$K_{52} = K_{31}^2 + K_{13}^3$$

$$K_{53} = K_{12}^3$$

$$K_{54} = K_{32}^2$$

$$K_{55} = K_{33}^2 + K_{11}^3 + K_{22}^4$$

$$K_{56} = K_{21}^4$$

$$K_{61} = 0$$

$$K_{62} = 0$$

$$K_{63} = 0$$

$$K_{64} = K^4_{13}$$

$$K_{65} = K^4_{12}$$

$$K_{66} = K^4_{11}$$

This information can easily be obtained by using the Boolean connectivity matrix for this particular problem as presented below.

$$[B] = \begin{bmatrix} 1 & 2 & 3 \\ 2 & 4 & 5 \\ 5 & 3 & 2 \\ 6 & 5 & 4 \end{bmatrix}$$

The Boolean connectivity matrix presents in a very easy form the connection between the local nodes for each element and the global nodes and from this inter-connectivity, the global matrix is easily derived. Study Fig. 4.3 and the construction of this Boolean matrix. This matrix is particularly useful in writing computer codes for matrix assembly. The numbers represent the numbers in the global matrix while the position of these numbers in the matrix represents the local values that would occupy the numbers on the global matrix. Try to reconcile the use of this connectivity matrix with the values presented in the connectivity values above.

4.6.1 Calculating the Column Vector

In this Laplacian, the column vector is the flux:

$$\{F\} = \left\{ \begin{array}{c} F_1 \\ F_2 \\ -1 \\ F_4 \\ 0 \\ 1 \end{array} \right\}$$

The flux value at nodes 1, 2 and 4 are unknown and the nodal values u_3, u_5 and u_6 can be calculated by solving the matrix equation $[K]\{u\} = \{F\}$ and substituting u_1, u_2 and $u_4 = 5.0$. The solution gives $u_3 = 6.66$, $u_5 = 5.76$ and $u_6 = 6.38$, and $F_1 = -0.83$, $F_2 = -0.76$, and $F_4 = -0.69$. The negative flux values show flux going in negative y-direction.

4.7 One-Dimensional Fourth Order Differential Equation (Beam Bending Problem)

$$-\frac{\mathrm{d}^2}{\mathrm{d}x^2}\left(a\frac{\mathrm{d}^2u}{\mathrm{d}x^2}\right) - f = 0 \tag{4.61}$$

This example is brought forth here to introduce the reader to an example of a C^1 continuous parameter function. For beam bending problem, $u =$ transverse deflective (positive upward) of the beam; $f =$ transverse distributed load acting downward.

The weak formulation is detailed in Chap. 2, Example 2.5.

However, remember in our section on parameter functions and interpolation functions we did talk about compatibility and completeness requirements for deriving shape functions for different problem statements. In this example, we cannot use a C^0 continuous shape function because this is a C^1 continuous equation. The interpolation function will therefore be a cubic polynomial despite it being a two-noded domain because there are four nodal variables for the beam element.

The interpolation function therefore takes the form for deflection as

$$\tilde{u} = c_1 + c_2 x + c_3 x^2 + c_4 x^3 \tag{4.62}$$

and the slope

$$(\mathrm{d}u/\mathrm{d}x) = \psi(x) = c_2 + 2c_3 x + 3c_4 x^2 \tag{4.63}$$

This Function (4.62) satisfies both the compatibility and completeness requirement for the problem solution.

$$0 = \int_{i}^{x_{i+1}} a\left[\frac{\mathrm{d}^2w}{\mathrm{d}x^2}\frac{\mathrm{d}^2u}{\mathrm{d}x^2} - wf\right]\mathrm{d}x + \left[w\frac{\mathrm{d}}{\mathrm{d}x}\left(a\frac{\mathrm{d}^2u}{\mathrm{d}x^2}\right) - \frac{\mathrm{d}w}{\mathrm{d}x}a\frac{\mathrm{d}^2u}{\mathrm{d}x^2}\right]_{x}^{x_{i+1}}$$

Primary variables are u and $\mathrm{d}u/\mathrm{d}x$.

The secondary variable is

$$\frac{\mathrm{d}}{\mathrm{d}x}\left[a\frac{\mathrm{d}^2u}{\mathrm{d}x^2}\right],$$

Deflection and slope at both beam ends of length l can be computed as;

$$u(0) = c_1 = u_1$$
$$\psi(0) = c_2 = \psi_1$$
$$u(l) = c_1 + c_2 l + c_3 l^2 + c_4 l^3 = u_2 \tag{4.64}$$
$$\psi(l) = c_2 + 2c_3 l + 3c_4 l^2 = \psi_2$$

Solving (4.64) in terms of the nodal variables gives the shape functions

$$h_1(x) = 1 - \frac{3x^2}{l^2} + \frac{2x^3}{l^3}$$

$$h_2(x) = x - \frac{2x^2}{l} + \frac{x^3}{l^2}$$

$$h_3(x) = \frac{3x^2}{l^2} - \frac{2x^3}{l^3} \tag{4.65}$$

$$h_4(x) = -\frac{x^2}{l} + \frac{x^3}{l^2}$$

Applying the approximation and weighting functions to (4.61)gives the finite element coefficient matrices $[K^e]$ and $\{f^e\}$:

$$[K^e] = \frac{2a}{l^3} \begin{bmatrix} 6 & 3l & -6 & -3l \\ 3l & 2l^2 & -3l & l^2 \\ -6 & -3l & 6 & -3l \\ 3l & l^2 & -3l & 2l^2 \end{bmatrix}$$

$$[f^e] = \frac{fl}{12} \begin{pmatrix} 6 \\ l \\ 6 \\ -l \end{pmatrix}$$

where $a = EI$ for beam bending.

4.8 The Use of Other Coordinate Systems in Derivation of Finite Element Equation

The natural coordinate system is very much used in finite element computation because of the ease of computing higher order element based matrices using simple integral methods and the Gauss–Legendre quadrature. This segment presents the application of some simple integral formulae for the length, area and volume coordinates. The calculation of the stiffness matrix still has to be through the Jacobian route outlined in Sect. 3.7 on isoparametric elements.

4.8.1 Length Coordinates

Let us derive the force vector of Eq. 4.7 using length coordinate shape functions

$$\int_{x_i}^{x_{i+1}} f\begin{pmatrix} h_1 \\ h_2 \end{pmatrix} dx = \int_l f\begin{bmatrix} L_1 \\ L_2 \end{bmatrix} dl = f \begin{bmatrix} \dfrac{1!0!}{(1+0+1)!} \\[2ex] \dfrac{0!1!}{(0+1+1)!} \end{bmatrix} = f\frac{l}{2}\begin{bmatrix} 1 \\ 1 \end{bmatrix}.$$

This gives us the force vector. Compare this solution with Eq. 4.8.

4.8.2 Area Coordinates

In this example we shall derive the first portion of the force vector matrix for the second order equation in Eq. 4.20

$$[F]^e = \int_{\Omega^e} \begin{Bmatrix} h_1 \\ h_2 \\ h_3 \end{Bmatrix} Q\, d\Omega = \int_A Q \begin{bmatrix} L_1 \\ L_2 \\ L_3 \end{bmatrix} dA = Q \begin{bmatrix} \dfrac{1!0!0!}{(1+0+0+2)!} \\[2ex] \dfrac{0!+1!+0!}{(0+1+0+2)!} \\[2ex] \dfrac{0!0!1!}{(0+0+1+2)!} \end{bmatrix} 2A = \frac{QA}{3}\begin{bmatrix} 1 \\ 1 \\ 1 \end{bmatrix}$$

4.8.3 Volume Coordinates

We shall calculate the force matrix for Eq. 4.36

$$\int_\Omega wQ\, d\Omega = \int_\Omega \begin{Bmatrix} h_1 \\ h_2 \\ h_3 \\ h_4 \end{Bmatrix} Q\, d\Omega = \int_V Q \begin{bmatrix} L_1 \\ L_2 \\ L_3 \\ L_4 \end{bmatrix} dV = Q \begin{bmatrix} \dfrac{1!0!0!0!}{(1+0+0+0+3)!} \\[2ex] \dfrac{0!1!0!0!}{(0+1+0+0+3)!} \\[2ex] \dfrac{0!0!1!0!}{(0+0+1+0+3)!} \\[2ex] \dfrac{0!0!0!1!}{(0+0+0+1+3)!} \end{bmatrix} 6V$$

$$= \frac{QV}{4}\begin{bmatrix} 1 \\ 1 \\ 1 \\ 1 \end{bmatrix}$$

These text [3, 4] provide further insight into mapping and working in other coordinate systems.

4.9 Exercises

4.1. Derive Eq. 4.8 from (4.7)

4.2. Solve for the Neumann boundary condition of Eq. 4.39 by using Eq. 4.48a

4.3. Check whether the interpolation function expressed in Eq. 4.62 satisfies both the compatibility and completeness requirement for Eq. 4.61

4.4. For a three meshed bar, derive the weak formulation, and the finite element equations for unsteady heat transfer in the bar

$$\frac{\partial T}{\partial t} - \frac{\partial}{\partial x}\left(a\frac{\partial T}{\partial x}\right) = f \quad 0 < x < 1$$

subject to the boundary conditions; $T(0) = T_0$; and

$$a\frac{\partial T}{\partial x} = -h_c(T - T^\infty) - \hat{q}$$

on Γ. Take a to be independent of x

4.5. Using a three element mesh, derive the weak formulation and hence the finite element equations for

$$\frac{\partial}{\partial x}\left(a\frac{\partial T}{\partial x}\right) = f \quad 0 < x < 1$$

Subject to the boundary conditions; $T(0) = T_0$; and

$$a\frac{\partial T}{\partial x} = -h_c(T - T^\infty) - \hat{q}$$

on Γ.

Take a to be independent of x.

References

1. Carnahan B, Luther HA, Wilkes JO (1969) Applied numerical methods. Wiley, New York
2. Yang WY, Cao W, Chung TS, Morris J (2005) Applied numerical methods using MATLAB®. Wiley, New York
3. Kardestuncer H (1987) Basic steps in the finite element method. In: Kardestuncer H, Norrie D (eds) Finite element handbook. McGraw-Hill, New York
4. Becker EB, Care GF, Oden JT (1981) Finite elements: an introduction, vol 1. Prentice-Hall, New Jersey

Chapter 5
Steps to Modeling Using PDEtoolbox^{TM} Graphics Interface

5.1 Engineering and Modeling

Many physical processes find their mathematical models expressed as partial differential equations. Designing for new products and processes need a great deal of modeling to understand the old processes and the proposed new design. Partial differential equations (PDEs) and their solutions are applicable to many engineering problems. This chapter sets the ground for a modeling journey using the PDEtoolbox graphics user interface (GUI).

5.2 Steps for Modeling with the PDEtoolbox

The following steps will guide the reader step-by-step through the modeling process. By the time the reader gains mastery, he/she can delve into modeling zones never explored before. We will outline these steps and build up on each in different segments of this chapter. The steps are:

1. derive the defining(governing) PDE for your problem, the domain and boundary conditions (and initial conditions if problem is time dependent);
2. open MATLAB and PDEtoolbox
3. specify the application you want by using the *Options → Applications* menu or the pull-down menu window;
4. draw the geometry(domain) of the problem on the graphics interface(GUI) using the *Options* and *Draw* menus;
5. set the applicable PDE and the initial and boundary conditions;
6. mesh the domain and refine mesh as applicable using adaptive or uniform meshing.
7. solve the PDE using the *Solve → SolvePDE* menu.
8. explore other solutions using the *Plot → Plot Parameters…* menu.

O. Oluwole, *Finite Element Modeling for Materials Engineers Using MATLAB®*, DOI: 10.1007/978-0-85729-661-0_5, © Springer-Verlag London Limited 2011

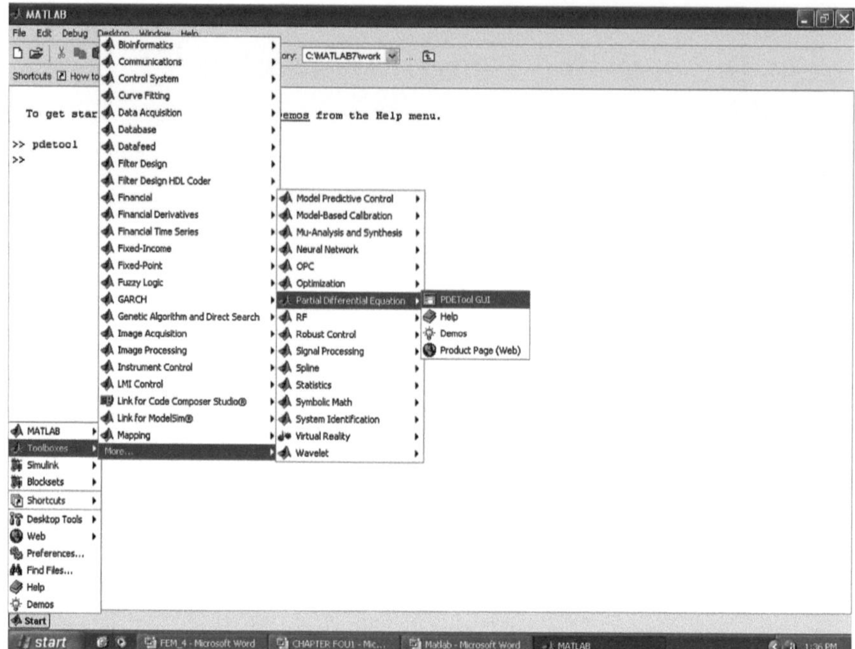

Fig. 5.1 Starting MATLAB 'PDEtool' GUI

Let us now go detailed into the outlined steps.

5.2.1 Starting the MATLAB PDEtool GUI

Start MATLAB by double-clicking on the desktop MATLAB icon. Type *PDETool* at the MATLAB prompt to start the GUI. Alternatively, as shown in Fig. 5.1 enter the graphic user interface through the MATLAB *start* button at the bottom left hand corner. Figure 5.2 shows the PDEGUI.

5.2.2 Specifying the Application Type

In the *Options → Application* menu (Fig. 5.3) or from the drop-down menu on the top right, select application type; heat transfer, diffusion, structural mechanics with plane strain, etc. Note that this selection must be made before specifying boundary conditions and PDE coefficients. Boundary values and PDE coefficients change if

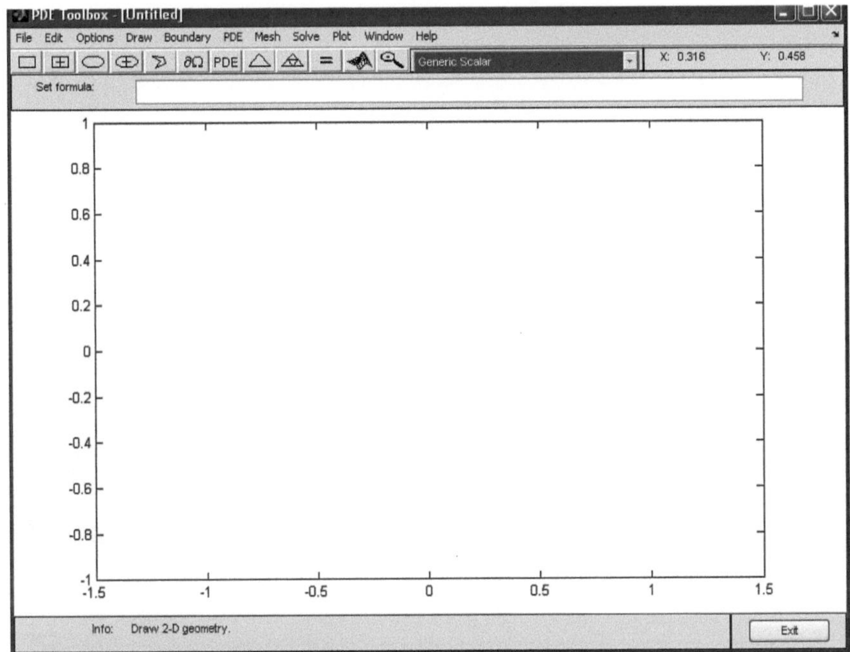

Fig. 5.2 The MATLAB PDEtoolGUI

Application selection is made after boundary conditions and PDE coefficients are set leading to errors in computation if changed values are not corrected.

5.2.3 *Drawing the Problem Geometry*

Draw the geometry, i.e., domain of the problem on the GUI using the *Options* and *Draw* menus. Use the *Options* → *Axes Limits…*menu to set displayed 2-D axes and *Options* → *Grid, Options* → *Grid Spacing…* and *Options* → *Snap* menus to automatically snap to grid points (Fig. 5.4). By drawing rectangles, ellipses or polygons, the domain can be built up by addition or subtraction from the *set formula* window (Figs. 5.5 and 5.6), e.g., SQ1 + C1 or SQ1 − C1.

Also, draw buttons in the tool bar just below the top menu bar can be used to draw the appropriate object by clicking on the appropriate icon. Click on the draw space and drag to draw the object.

Note carefully that it is important to choose convenient axes limits, preferably on a 1:1 scale with geometry dimensions.

Using *Boundary* → *Boundary Mode* and *Boundary* → *Remove Subdomain Border*, unwanted subdomain borders can be removed by selecting the unwanted border and clicking *Remove Subdomain Border*.

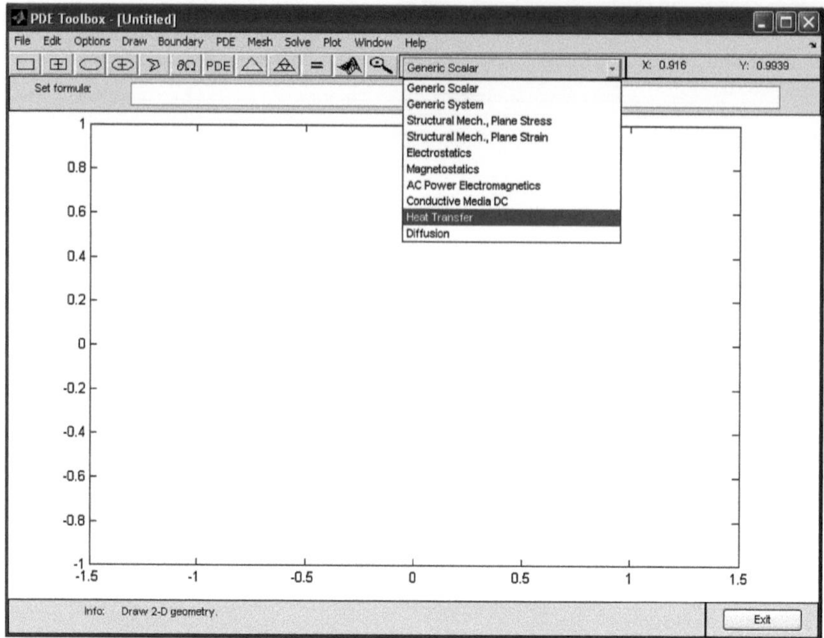

Fig. 5.3 Selecting the application type

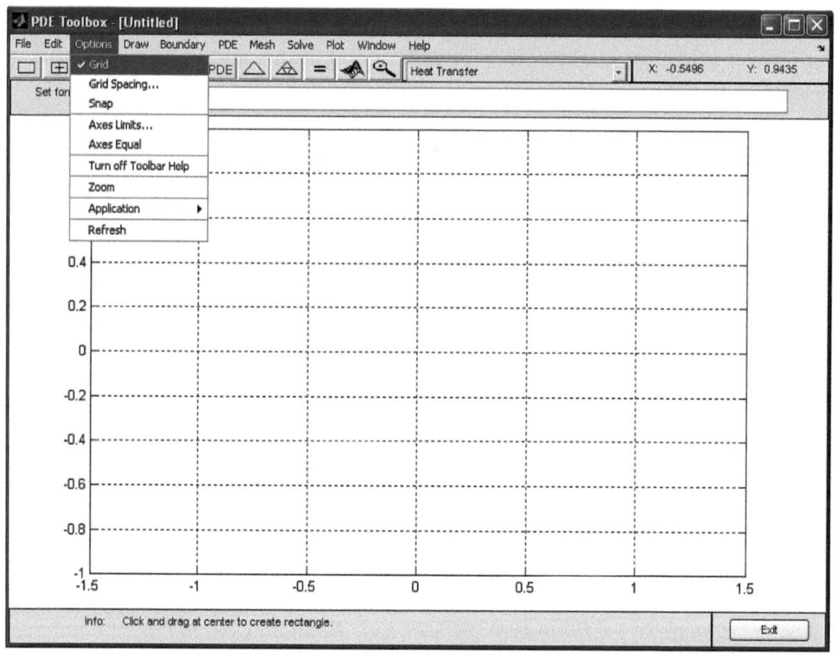

Fig. 5.4 Grid spacing before drawing

Fig. 5.5 Drawing the problem domain using the Draw menu

5.2.4 Specifying the PDE

Using the *PDE → PDE mode,* and *PDE → PDE specification…* (Figs. 5.7 and 5.8) the PDE type can be selected and the PDE coefficients can be entered. In 'multiple' domained problem, you must select the domain whose values are being entered, e.g., heat transfer in a molten steel casting enclosed in sand mould. The mould must be selected for entry of its parabolic PDE values while different parabolic PDE values must be entered for the molten metal.

5.2.5 Specifying Boundary Conditions

Enter boundary mode by clicking on the $\partial\Omega$ button or selecting *Boundary → Boundary Mode* (Figs. 5.9 and 5.10). Boundary conditions can be selected when in boundary mode by double-clicking on the desired boundary and filling the boundary conditions form (Fig. 5.11). Alternatively, single-click on the desired boundary and click *Boundary → Specify Boundary Conditions…* to specify boundary conditions. If several boundaries have the same conditions, they can be selected by *shift*-clicking on the desired boundaries. While in boundary mode all

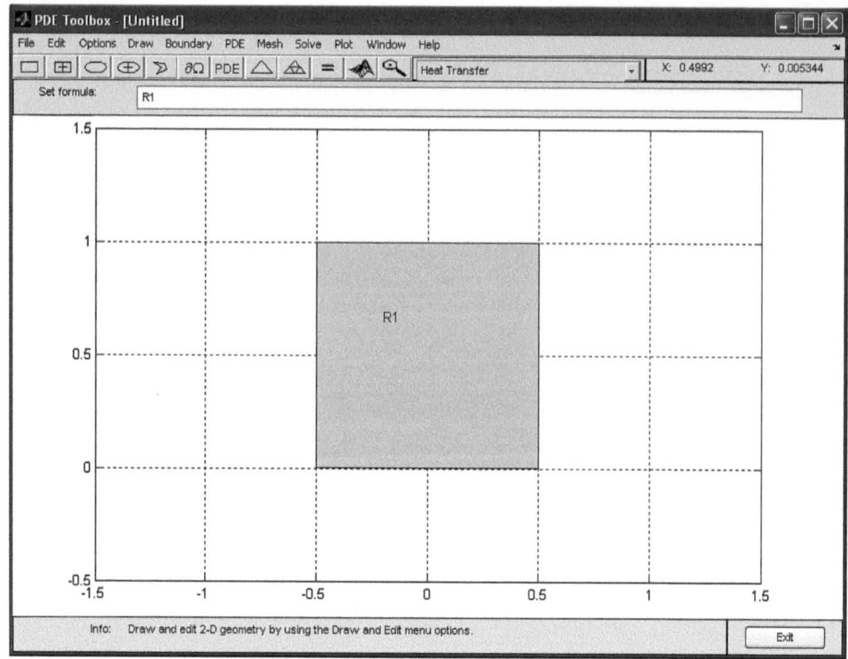

Fig. 5.6 A square domain

boundaries can be selected by using *Edit → Select All* and *Boundary → Specify Boundary Conditions…* to specify boundary conditions for this group of segments. Note that as you use the PDEtool GUI, you will find the most convenient way of getting around the use of this facility.

5.2.6 Meshing the Domain and Mesh Refinement

Using *Mesh → Parameters,* initial mesh parameters, jiggle mesh parameters and mesh refinement method can be selected (Figs. 5.12 and 5.13). For coarsest mesh *inf* can be input in the form by clicking on the triangle button or selecting *Mesh → Initialize Mesh,* initial mesh can be generated.

For non-transient problems adaptive mode of mesh refinement is an option and can be selected from *Solve → Solve Parameters* menu (Figs. 5.14 and 5.15). Here, the desired maximum number of triangles or maximum number of refinement steps can be entered in the form.

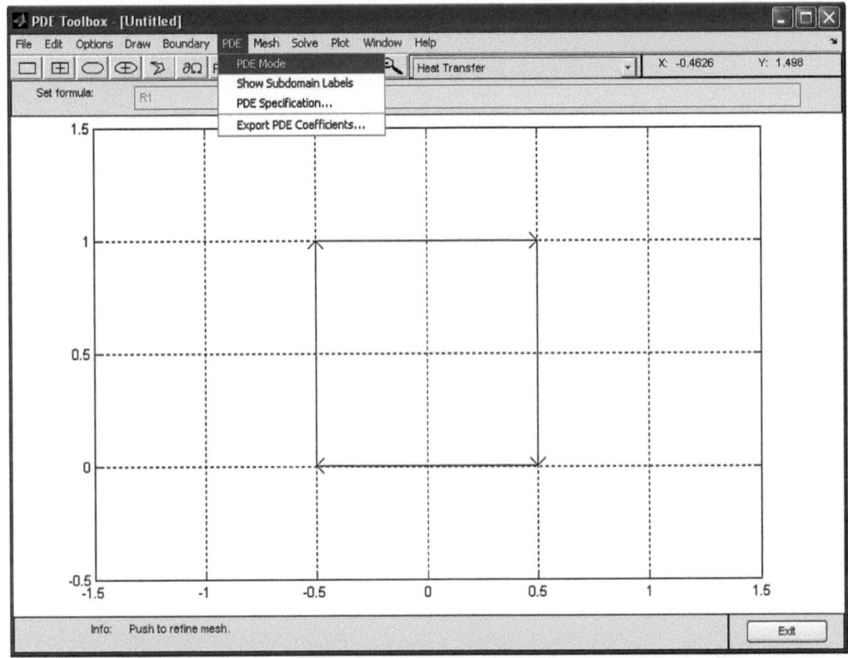

Fig. 5.7 The PDE menu for specifying the PDE type and coefficients

Fig. 5.8 PDE specification form for heat transfer application

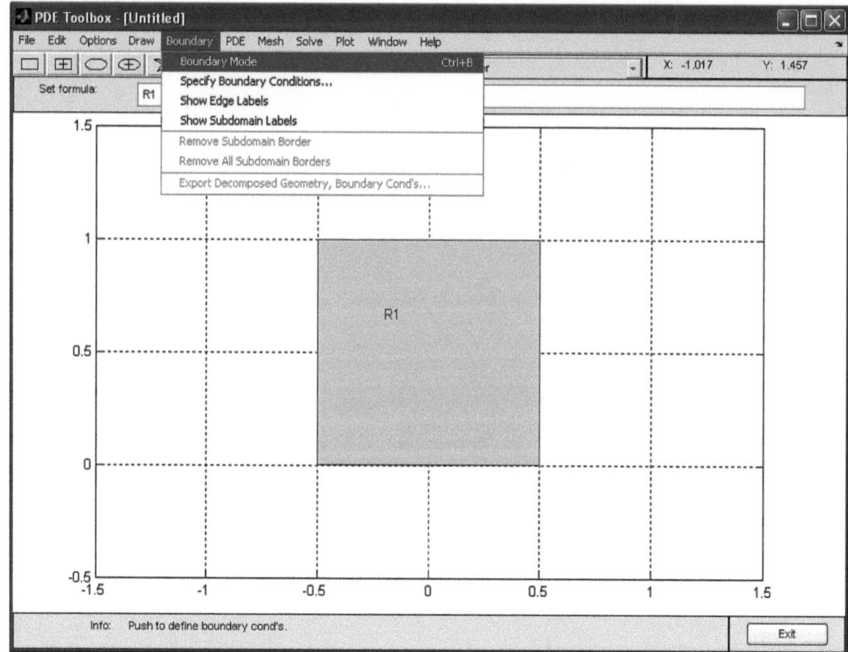

Fig. 5.9 Using the boundary mode

5.2.7 Specifying Initial Conditions for Transient Problems

If the problem is transient, the *Solve → Parameters…* menu (Figs. 5.14 and 5.16) is used for setting the time duration and increments. The variable $u(t0)$ represents the initial conditions on the domain at $t = 0$. *Time:0:?* represents $t = 0$ and $t = $ final time in seconds (Fig. 5.13).

5.2.8 Solving the PDE

Solving the PDE involves plotting the result on the GUI or on a separate pop-up graph in 3D. This means that the properties to be plotted must be selected. By selecting *Plot → Parameters,* or clicking on the mesh button, the plot selection form pops up and the plot properties can be selected (Figs. 5.17 and 5.18). For *Colormap options, jet* or *cool* is recommended. Plotting in *x–y grid, show mesh* and *Height(3D plot)* are also recommended but user can choose appropriately as desired after familiarizing with tools. After selection of plot properties, click on the plot button on the plot parameters form or click on *Solve → Solve PDE…menu* or press the $=$ button.

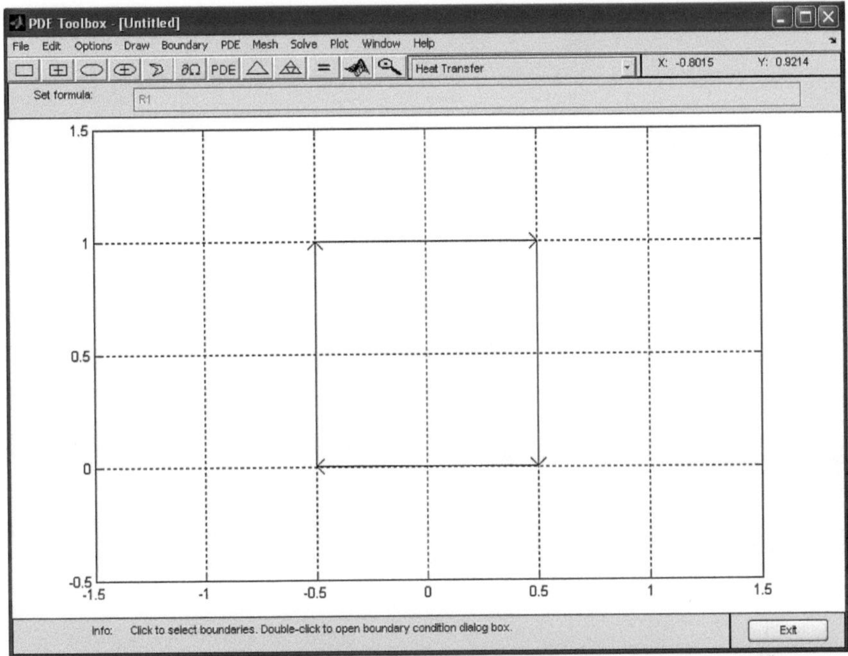

Fig. 5.10 The square domain in boundary mode

Fig. 5.11 The boundary condition form for heat-transfer application

5.2.9 Extracting Values from Plots

Values of plotted property at desired points in the domain can be obtained by clicking on the selected points in the domain, with the values displaced on the status bar at bottom of the GUI. Exact solutions at these points can be found by using the *Mesh → Export Mesh* and *Solve → Export Solution* menus to export the mesh parameters and the solution parameters to the MATLAB worksheet. The

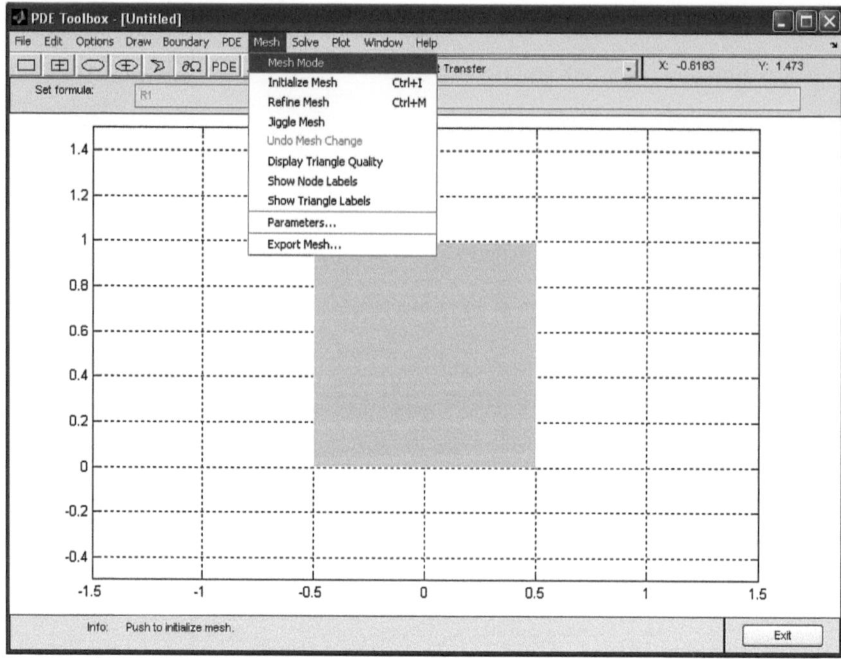

Fig. 5.12 Meshing mode menu showing the mesh options

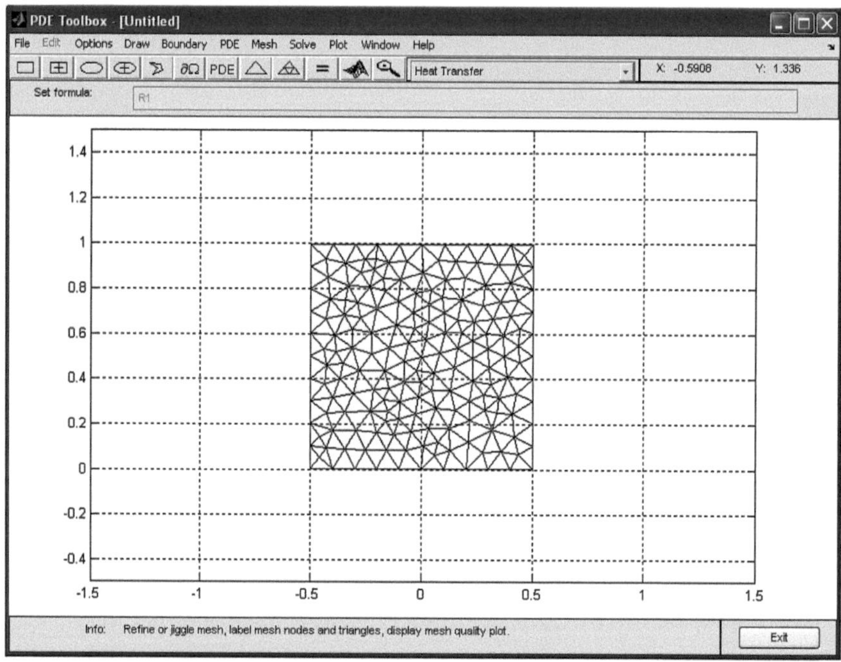

Fig. 5.13 A triangular meshed domain

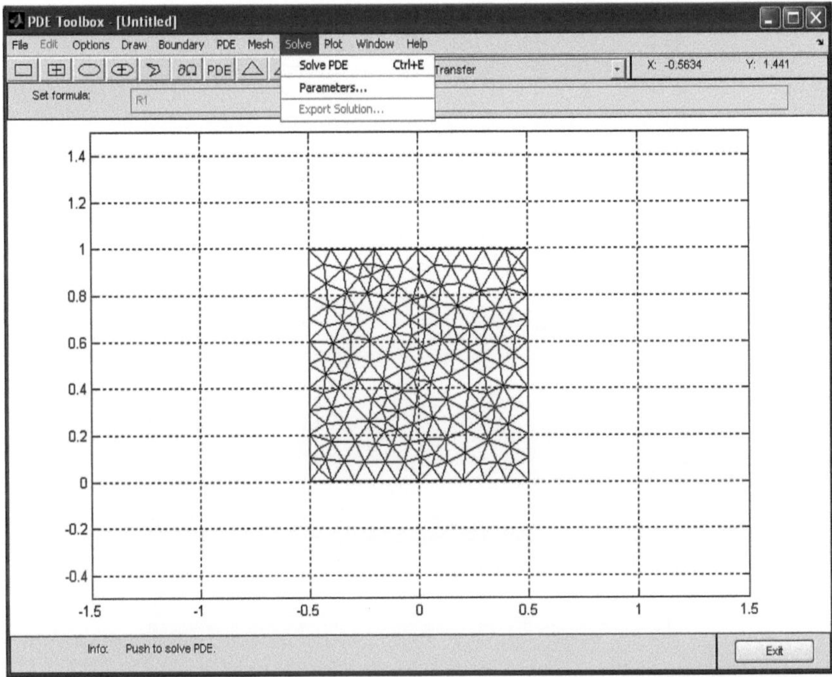

Fig. 5.14 Selecting the solve option

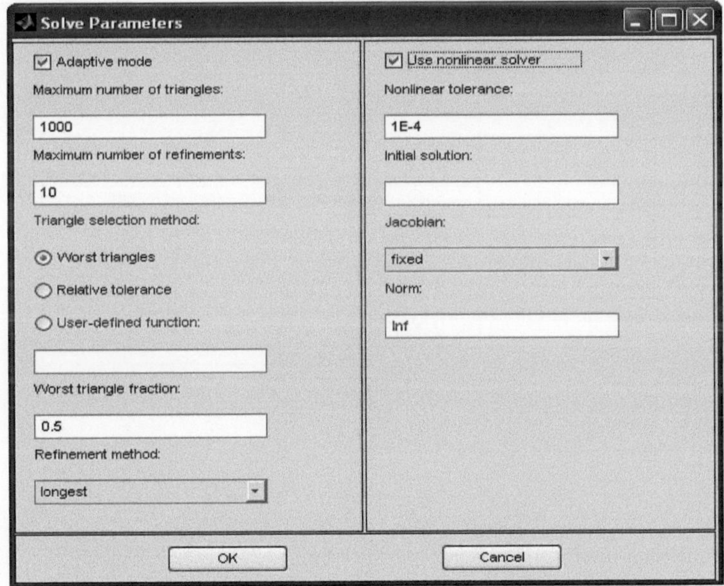

Fig. 5.15 Solve parameters form for non-transient problems

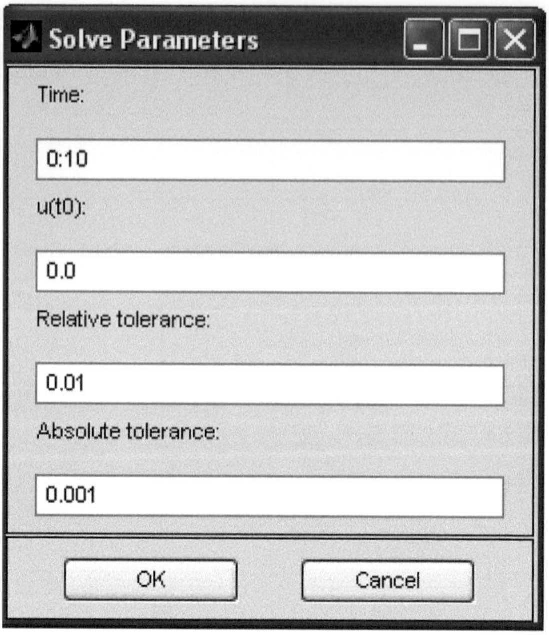

Fig. 5.16 Solve parameters input form for transient heat-transfer application

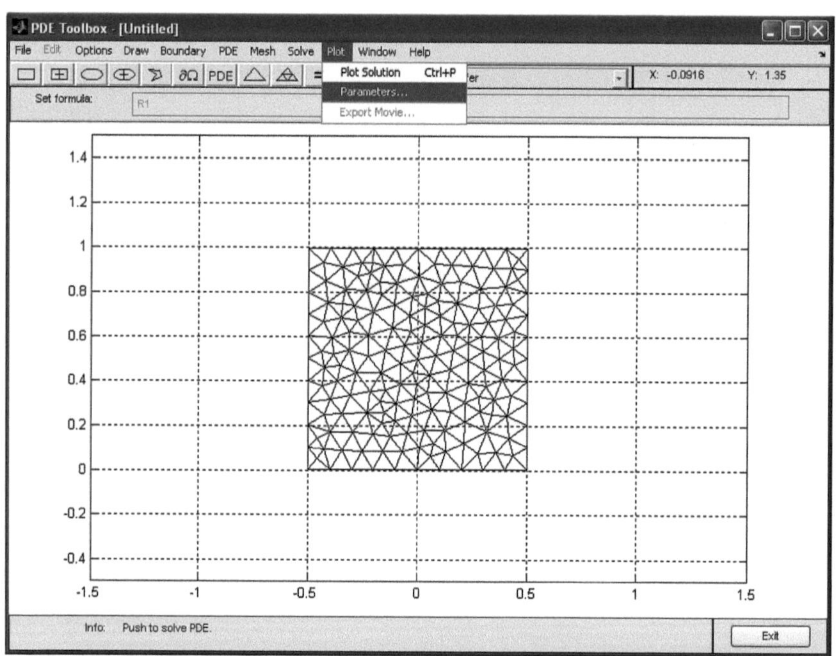

Fig. 5.17 Selecting the plot parameters

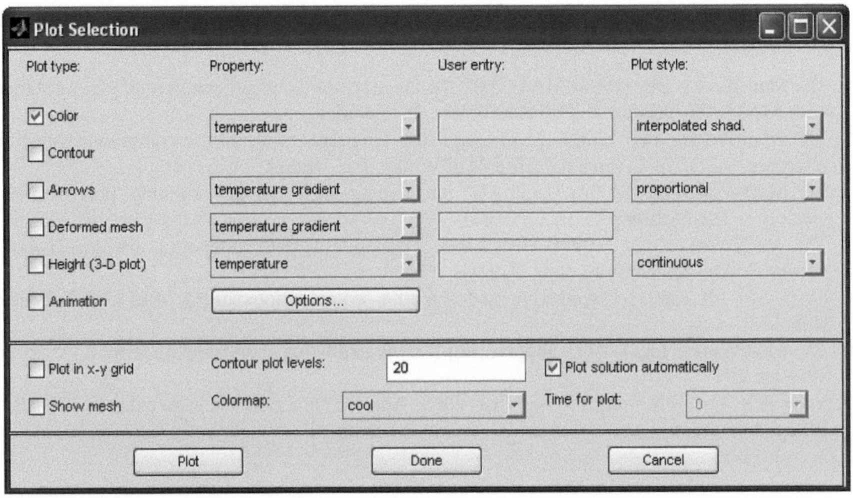

Fig. 5.18 The plot selection form

desired value $u(x,y)$ can then be obtained from the MATLAB worksheet by evaluating `uxy = tri2grid(p,t,u,x,y)` from the MATLAB prompt inserting specific values of `x` and `y` into the `x,y` bracketed with `p,t,u`. The mesh triangle number `tn` can also be obtained using `[uxy,tn] = tri2grid(p,t,u,x,y)`. The vertices of the triangle can be found by evaluating `p(:,t(1:3,tn))` from the MATLAB prompt. Note that other properties of your plot can be extracted as well, e.g. parameters involved in your domain geometry or the boundary conditions or PDE or even parameters involved in your graph. Go to the appropriate pull-down menu and click the desired property *Export…*

Some useful texts for the reader are listed in the Refs. [1–7].

5.3 Exercises

5.1. Use the draw icons situated below the top menu bar to draw a circle of radius 2, rectangle of sides $0 \le x \le 4$ and $0 \le y \le 2$, and square with $0 \le x \le 2$ and $0 \le y \le 2$.

5.2. Draw a square of sides 4 cm with a square hole of sides 2 cm.

5.3. Draw three circles of radius $= 1$ inside a rectangle of sides $0 \le x \le 6$ and $0 \le y \le 4$.

5.4. Select one of the circles drawn in Exercise 4.3 by clicking on it and delete.

5.5. Re-label one of the remaining two circles in Exercise 4.3 by double-clicking on it to open the object dialogue box.

5.6. Draw an equilateral triangle of sides $= 1$. Open the rotate box from the draw menu and rotate $90°$.

References

1. The MathWorks, Inc. (1998) MATLAB® the language of technical computing: getting started with MATLAB, version 5. The MathWorks, Inc., Natick
2. The MathWorks, Inc. (2004) MATLAB® the language of technical computing: function reference, vol 1: A–E version 7. The MathWorks, Inc., Natick
3. The MathWorks, Inc. (2004) MATLAB® the language of technical computing: programming, version 7. The MathWorks, Inc., Natick
4. The MathWorks, Inc. (1999) SIMULINK® dynamic system simulation for MATLAB®, version 3. The MathWorks, Inc., Natick
5. Lyshevski SE (2003) Engineering and scientific computations using MATLAB®. Wiley, Hoboken
6. The MathWorks, Inc. (2011) Support. http://www.mathworks.com/support/?BB=1. Accessed 19th Jan 2011
7. Yang WY, Cao W, Chung TS, Morris J (2005) Applied numerical methods using MATLAB®. Wiley, New York

Chapter 6
Application of PDEtoolbox^{TM} to Heat Transfer Problems

6.1 Setting-Up the GUI for Heat Transfer Problems

Basic procedures required in heat transfer problems are setting the application mode, drawing the domain geometry, imposing the boundary conditions, the PDE specification and plot specifications. Great care need to be taken to make sure that all details in the governing equation are taken care of. We will look at these procedures further.

Application mode. Set menu *Options* → *Application* to heat transfer

Boundary conditions. For Dirichlet condition expressed as $h*T = r$, set 'h' to unity and 'r' to the temperature desired.

For Neumann condition expressed as $n*k*\text{grad}(T) + q*T = g$, set '$q$' to zero and '$g$' to negative of desired heat flux (or zero for adiabatic condition). n is the outward unit normal vector.

For convective boundary condition, set 'q' to convective heat transfer coefficient value and 'g' to product of heat transfer coefficient and ambient temperature.

PDE specification. The elliptic Fourier equation expressed as $-\text{div}(k*\text{grad}(T)) = Q + h*(T_{\text{ext}} - T)$ over a domain Ω is for steady-state situations, while the parabolic equation, $rho*C*T' - \text{div}(k*\text{grad}(T)) = Q + h*(T_{\text{ext}} - T)$ describes transient problems.

6.2 Example Problems on Heat Transfer in Materials Engineering

Some sample problems in Material Engineering are presented in this segment to familiarize the reader with how to use PDEtoolbox in materials modeling. Useful texts are listed in the Refs. [1–5].

Note that all the codes in the sample problems given in this book can be accessed on http://extras.springer.com/. To follow the modeling procedure, the

O. Oluwole, *Finite Element Modeling for Materials Engineers Using MATLAB®*,
DOI: 10.1007/978-0-85729-661-0_6, © Springer-Verlag London Limited 2011

reader can actually copy portions of the code to be followed on to a new M-FILE. A new M-file can be opened by clicking on the new M-file icon on the MATLAB workspace just below the menu bar. Run the program by clicking on the Debug menu and double-clicking save and run on the pull-down menu option. This will enable you to follow the solution procedure in an all-color mode. Other portions can be added as the reader goes along. For example, the code dirichlet.m listed below (see Sect. 6.2.2.1); to follow the geometry construction, copy the codes up to geometry description and run. Later on to follow how the boundary conditions were set up, add the codes for boundary conditions and run again. Then add codes for mesh generation…, then pde coefficients, etc. until you get codes to solve pde. The reader can follow step-by-step in this way. Running all the codes will give the final solution in the domain geometry.

6.2.1 Steady-State Heat Transfer

A metal 0.50 m by 0.01 m ($K = 73$ W/m °C) has one end kept at uniform temperature of 400°C and the other at 25°C. The sides are insulated.

Using PDEtoolbox, find the steady-state temperature distribution in the metal. The solution to this problem can be accessed in the METAL.m file.

Solution steps:
1. Open the PDEtoolGUI and specify heat transfer application.
2. Choose the right scales for your domain, grid and snap from Options menu (see Chap. 5 on geometry drawing). Draw your domain using the Draw icons or the draw menu. You could choose axes limits of [0 0.6] and [0 0.02] for both x and y axes in the menu Options → axes limits and then snap to 0.5 by 0.01 or using the object dialogue box set Left-Bottom-Width-Height in the dialogue box to 0-0-0.5-0.01, respectively.
3. Select the PDE. First click on PDE mode in the PDE drop down menu list. Then click on the geometry to open up the PDE specification box. The elliptic PDE is applicable here for steady-state situation:

   - div(k*grad(T)) = Q + h*(T_{ext} − T) where T is temperature, h is the convective heat transfer coefficient, T_{ext} is the ambient temperature. Your knowledge of finite element equation derivation obtained from Chaps. 2 to 4 comes in useful here. From the problem formulation, we have only dirichlet conditions involved here, the Neumann conditions are zero, no heat source and no convective conditions. Insert values into the PDE specification form. $Q = 0$, $h = 0$, $T_{ext} = 0$.

4. Put in the boundary conditions, the insulated sides are kept as Neumann's condition

$$\frac{dT}{dx} = 0 \text{ by keeping } q = 0 \text{ and } g = 0.$$

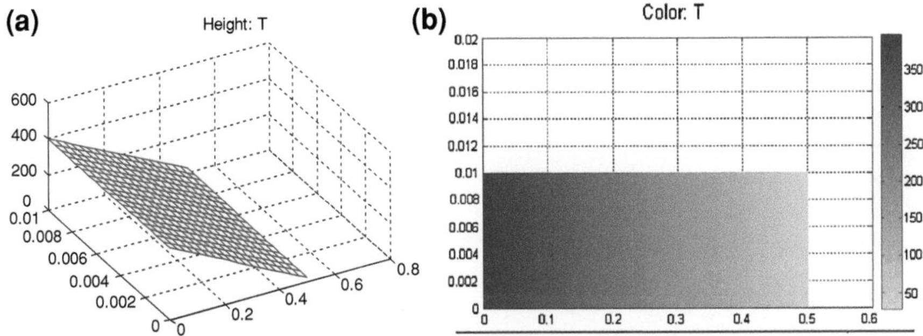

Fig. 6.1 Solution to Example 6.2.1 in **a** 3D plot mode, **b** 2D colour plot mode

The ends are dirichlet conditions with $T = 400$ and $T = 25$. Keep $h = 1$.

5. Mesh your domain and refine mesh. You may use the adaptive refinement method here by filling the solve $\rightarrow$ parameters form.
6. Select your plot parameters and press the plot button.

The solution is presented in Fig. 6.1.

6.2.2 Transient Problems (Heating and Cooling Problems)

6.2.2.1 Heating Problems

A metal plate 1 m by 1 m was initially at a temperature of 25°C. Suddenly the boundaries were kept at 250, 100, 500 and 25°C going anticlockwise from the top. The flat surface itself was experiencing convective heat transfer. Find the temperature distribution in (i) 10 s, (ii) 3000 s. Use the values; $Q = 0$, $K = 73.4$ W/m °C, $C_p = 1300$ J/kg/K, $\rho = 7680$ kg/m^3, $h = 28.39$ W/m^2 °C and $T_{ext} = 25$°C.

The solution to this problem can be accessed by running the *dirichlet* M-file.

Solution steps to 6.2.2.1(i). Open dirichlet.m file to follow this solution procedure:

1. Open the PDFtoolGUI and specify heat transfer application.
2. Choose the right scales for your domain. I suggest $1 = 1$ scale using *Options $\rightarrow$ Axes Limits…*; *Options $\rightarrow$ Grid* and *Options $\rightarrow$ Snap*. Draw your domain using the RECTANGLE button or Draw $\rightarrow$ Rectangle/Square.
3. Select the parabolic PDE which is applicable for transient heat transfer; $rho*C*T' - \text{div}(k*\text{grad}(T)) = Q + h*(T_{ext} - T)$ where *rho* is the density, C is the specific heat capacity at constant pressure, T' is the derivative of T with respect to time. Insert the values. $Q = 0, K = 73.4$ W/m °C, $C_p = 1300$ J/kg/K, $\rho = 7680$ kg/m^3, $h = 28.39$ W/m^2 °C and $T_{ext} = 25$°C.
4. Put in the boundary conditions (B.C's). All the B.C's are dirichlet conditions.
5. Mesh the domain and refine mesh.

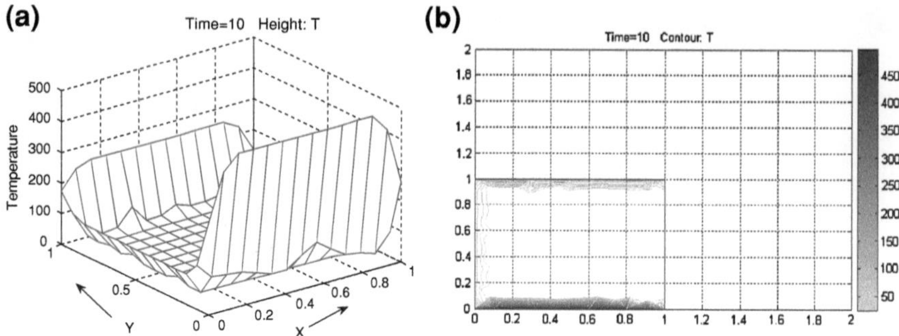

Fig. 6.2 a Solution to 6.2.2.1(i) in **a** 3D plot mode, **b** contour plot mode

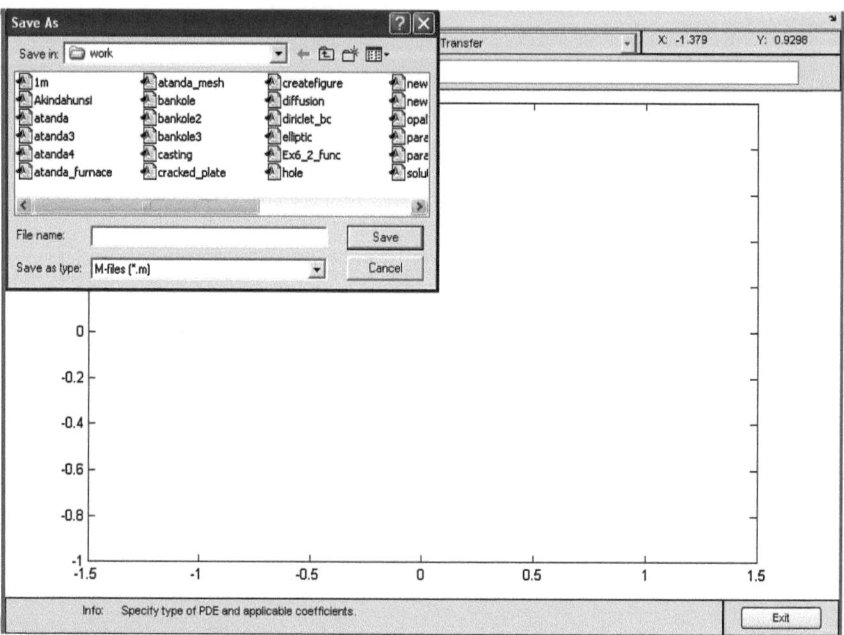

Fig. 6.3 Saving your work

6. Open the *solve parameters* form putting in the initial conditions $u(t0) = 25$ and
   final time $t = 10$. The solutions are presented in Fig. 6.2.

You can off the grid from the 'Options' menu as desired. From the 'File' menu
on the PDEToolbox GUI select 'save as' and save your work. Your work is saved
into the MATLAB work folder (Fig. 6.3). Your saved work is called an M-file.
Opening it reveals the codes for your work.

From the file menu of the MATLAB GUI (not the file menu of the PDEToolbox
GUI) the saved work can be opened as an M-file (Fig. 6.4). If the M-file is opened
from the PDEtoolGUI, the codes are executed.

```matlab
% Dirichlet.m file
function pdemodel
[pde_fig,ax]=pdeinit;
pdetool('appl_cb',9);
pdetool('snapon','on');
set(ax,'DataAspectRatio',[1 1.5 1]);
set(ax,'PlotBoxAspectRatio',[1 0.66666666666666663 1]);
set(ax,'XLim',[0 2]);
set(ax,'YLim',[0 2]);
set(ax,'XTickMode','auto');
set(ax,'YTickMode','auto');
pdetool('gridon','on');

% Geometry description:
pderect([0 1 1 0],'SQ1');
set(findobj(get(pde_fig,'Children'),'Tag','PDEEval'),'String','SQ1')

% Boundary conditions:
pdetool('changemode',0)
pdesetbd(4,...
'dir',...
1,...
'1',...
'100')
pdesetbd(3,...
'dir',...
1,...
'1',...
'500')
pdesetbd(2,...
'dir',...
1,...
'1',...
'25')
pdesetbd(1,...
'dir',...
1,...
'1',...
'250')

% Mesh generation:
setappdata(pde_fig,'Hgrad',1.3);
setappdata(pde_fig,'refinemethod','regular');
pdetool('initmesh')
```

```
% PDE coefficients:
pdeseteq(2,...
'73.4',...
'28.39',...
'(0)+(28.39).*(25.0)',...
'(7680.0).*(1300)',...
'0:3000',...
'25.0',...
'0.0',...
'[0 100]')
setappdata(pde_fig,'currparam',...
['7680.0';...
'1300  ';...
'73.4  ';...
'0     ';...
'28.39 ';...
'25.0  '])

% Solve parameters:
setappdata(pde_fig,'solveparam',...
str2mat('0','1000','10','pdeadworst',...
'0.5','longest','0','1E-4','','fixed','Inf'))

% Plotflags and user data strings:
setappdata(pde_fig,'plotflags',[1 1 1 1 1 1 1 1 0 0 0 3001 0 1 0 0 0 1]);
setappdata(pde_fig,'colstring','');
setappdata(pde_fig,'arrowstring','');
setappdata(pde_fig,'deformstring','');
setappdata(pde_fig,'heightstring','');

% Solve PDE:
pdetool('solve')
```

Solution of 6.2.2.1(ii). Follow the same procedure as in the solution of 6.2.2.1(i) but insert $t = 3000$ s in the *solve parameters* form. The solution is presented in Fig. 6.5.

6.2.2.2 Cooling Problems

The same steel plate of Sect. 6.2.2.1 with same boundary conditions was initially at 900°C. Find temperature distribution after (i) 10 s in air; (ii) 3000 s in air; (iii) 3000 s quenching the sides alone to 25°C; and (iv) subjected to Neumann B.C $dT/dn = -500$ on all sides.

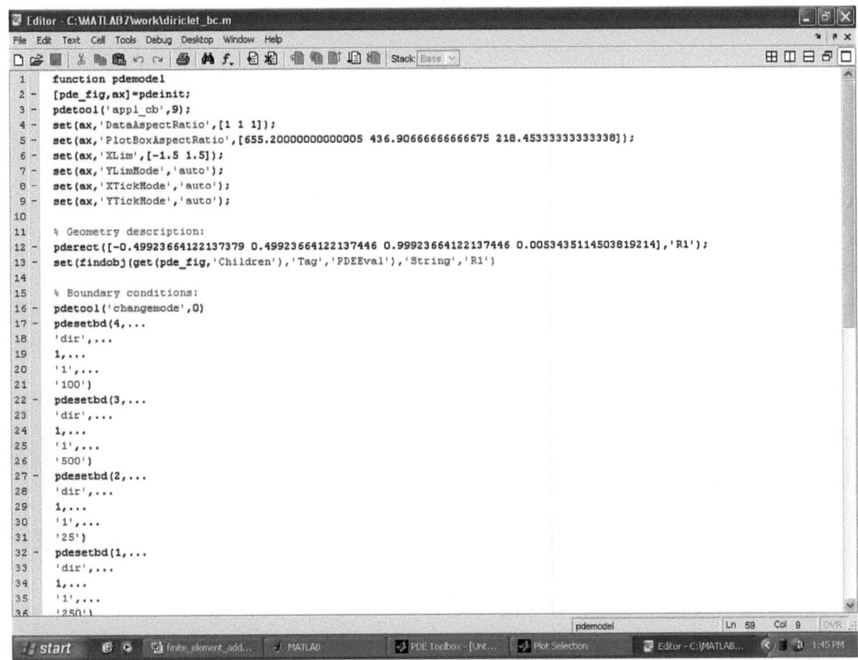

Fig. 6.4 The M-file shown under Editor-C:\ MATLAB7\work\dirichlet M file. The full contents of this file is listed below and can be run from the debug menu tool 'run'

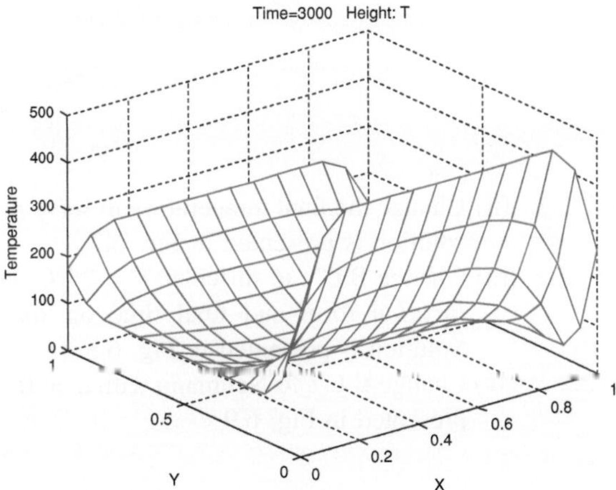

Fig. 6.5 Solution to 6.2.2.1(ii)

Solutions:

6.2.2.2.   (i) Apply the same procedure as in Problem 6.2.2.1 but in the solve parameters form input initial temperature as 900°C.

The solution is presented in Fig. 6.6. You will notice that the plate is cooling down because the initial temperature condition was higher than subsequent temperature the plate was subjected to.

Fig. 6.6 Solution to Problem
6.2.2.2(i)

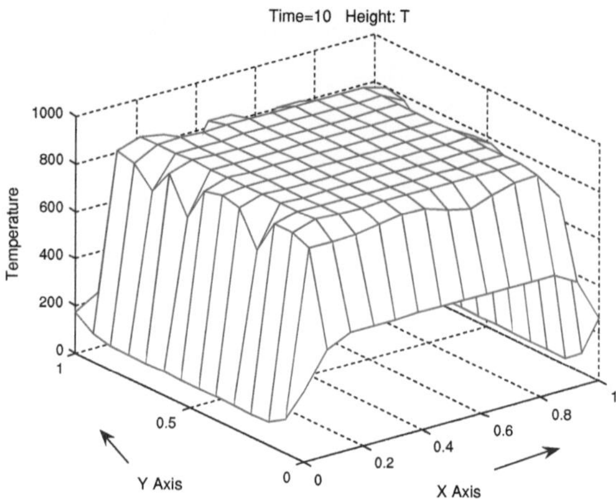

Fig. 6.7 Solution to Problem
6.2.2.2 (ii)

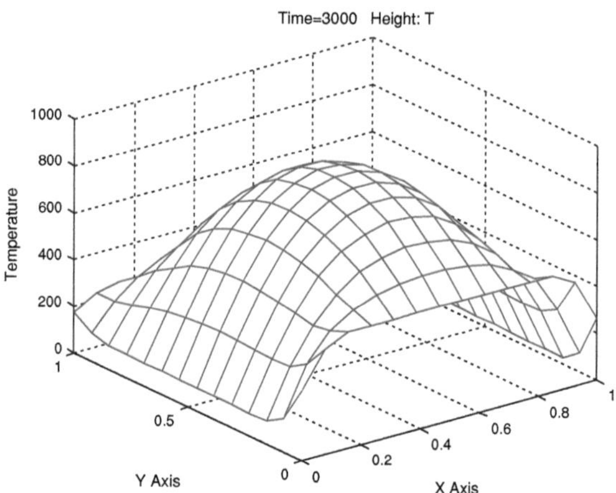

6.2.2.2. (ii) Change the final temperature in *solve parameters* form to 3000 s. The
solution is presented in Fig. 6.7.

6.2.2.2. (iii) Change B.C's to dirichlet $T = 25°C$ on all sides on *Boundary Specification* form and final time on *solve parameters* form to 3000.
Solution is presented in Fig. 6.8.

6.2.2.2. (iv) Change B.C's to Neumann with $q = 0$ and $g = -500$. The solution is
presented in Fig. 6.9.

6.2.3 Transient Problem (Heat Generation in a Tubular Furnace)

A tubular furnace has an internal diameter of 0.5 m with a heating element
embedded in a fireclay refractory ring of internal diameter of 0.5 m. A backup

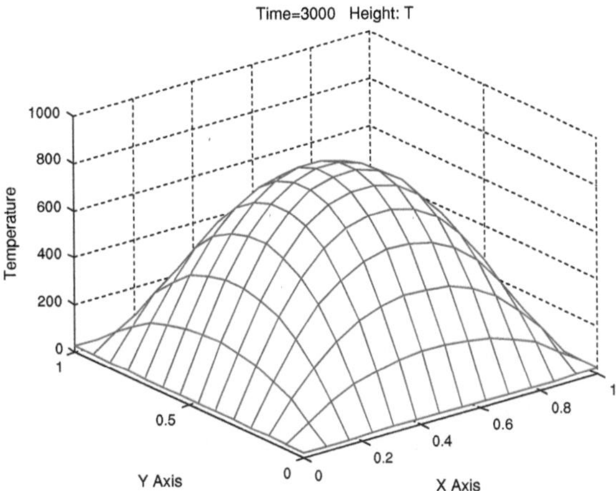

Fig. 6.8 Solution to Problem 6.2.2.2(iii)

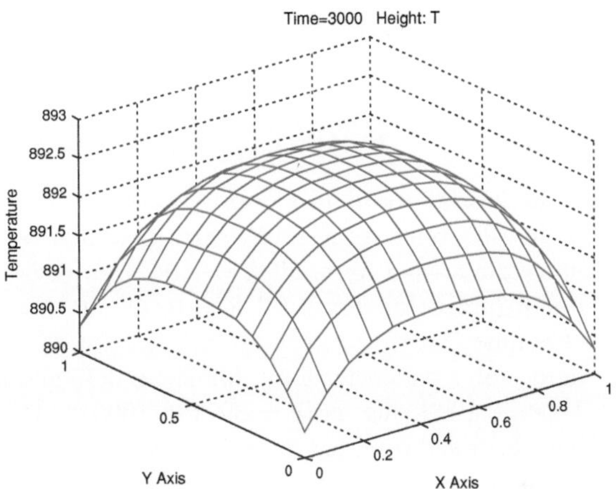

Fig. 6.9 Solution to Problem 6.2.2.2(iv) with Newmann BCs at $T = 3000$ s. We can see that the rate of cooling was much less than the side quenching situation from the temperature

refractory of silica bricks follows the fireclay ring with an internal diameter of 0.7 m. It is followed by a glass-wool insulation with an internal diameter of 1 m after which a steel shell of internal diameter of 1.1 m and external diameter of 1.2 m. The heating element has a power of 715 W and has capacity to heat up to 1200°C. Find the temperature distribution from the center of the furnace to the end of the glass-wool insulator in 30 s and 1000 s. if the temperature at the end of the glass wool is 30°C. Initial temperature was 30°C.

Data: K(firebrick) $= 0.15$ W/m °C; K(silicabrick) $= 0.6$ W/m °C; K(glass-wool) $= 0.04$ W/m °C; h(air) $= 28.39$ W/m^2 °C. Take the still air temperature in the furnace to obey the equation, Text $= 30 + t$.

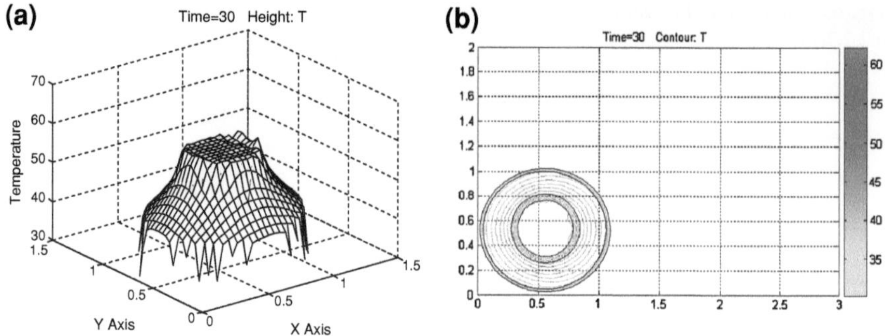

Fig. 6.10 **a** Temperature distribution in the tubular furnace after 30 s in 3D plot. **b** Temperature distribution in the tubular furnace after 30 s in 2D contour plot

Solution:
1. Draw the furnace leaving out the steel shell.
2. Fill the *PDE specification* and the *solve parameter* forms.

Keep *rho* and $C_{\text{p}} = 1$ to reduce the parabolic equation to

$$T' - \text{div}(k * \text{grad}(T)) = Q + h * (T_{\text{ext}} - T).$$

In the fireclay form, fill $k = 0.15$, $Q = 715$, $h = 28.39$, Text $= 30 + t$
In the silicabrick form, fill $k = 0.6$, $Q = 0$, $h = 0$, Text $= 0$
In the glass-wool form, fill $K = 0.04$, $Q = 0$, $h = 0$, Text $= 0$
In the furnace air form, fill $K = 0$, $Q = 0$, $h = 28.39$, Text $= 30 + t$ where
$t = $ time
In the solve parameter form, fill Time: 0:30 and $u(t0) = 30$ for 30 s and
Time: 0:1000 and $u(t0) = 30$ for 1000 s. The solutions are presented in
Figs. 6.10 and 6.11 and the M-file *transienttubular contains the codes.*

6.3 Exercises

6.1 Find the heat flux and temperature gradient for Example 6.1 using *METAL*.m file in the accompanying CD.

6.2 Find the temperature at $x = 0.20$ and $y = 0.004$ by clicking on the solution of Example 6.2.1 using the contour two-dimensional plot presented in METAL.m file.

6.3 Using example METAL.m file, find the temperature at $x = 0.20$ and $y = 0.004$ by exporting the mesh and solution parameters and using the expression `uxy = tri2grid(p,t,u,x,y)`. You should get a value of 250°C.

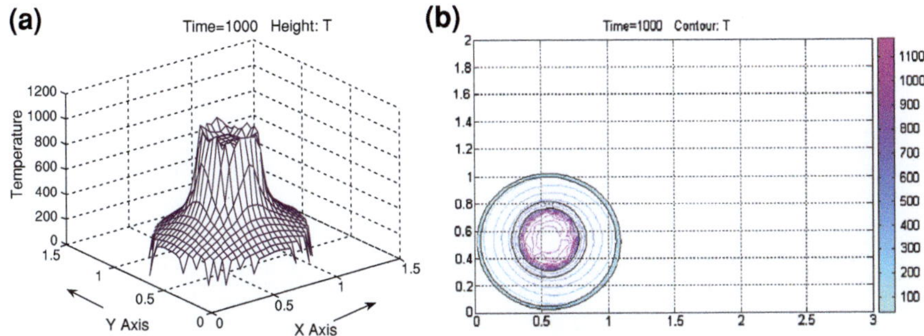

Fig. 6.11 a Temperature distribution in the tubular furnace after 1000 s in 3D plot. **b** Temperature distribution in the tubular furnace after 1000 s in 2D contour plot

6.4 Evaluate the triangle number (tn) of Exercise 6.3 and the vertices of the triangle.
   Solution: tn = 69; vertices $[x_1, y_1] = [0.2033, 0.005]$ $[x_2, y_2] = [0.1917, 0.0025]$ $[x_3, y_3] = [0.2033, 0.0025]$

6.5 Using *dirichlet* m file, find the heat flux and temperature gradients in Examples 6.2.2.1 and 6.2.2.2. Find the specific temperatures at $x = 0.2$, $y = 0.5$ using $[uxy] = $ tri2grid(p,t,u,x,y) as described in Chap. 5.

6.6 Solve the same problem in Sect. 6.2.3 for a steady-state situation with the furnace at 1200°C. Hint: Text = 1200°C inside furnace. PDE is elliptic. Solution is presented in *steadystatetubular M-file*.

6.7 Find the temperature gradients and heat flux in the tubular furnace using *steadystatetubular* and *transienttubular* M-files in the accompanying CD.

6.8 Extract the temperature values at the middle of the furnace for both transient at 1000°C and steady state and compare values. Solve for temperature distribution in the furnace for time situations of 50 and 100 s.

References

1. Holman JP (1990) Heat transfer, 7th edn. McGraw-Hill, New York
2. Kreith F, Bohn MS (1993) Principles of heat transfer, 5th edn. West Publishing, St Paul
3. Ozisik MN (1985) Heat transfer: a basic approach. McGraw-Hill, New York
4. Myers GE (1971) Analytical methods in conduction heat transfer. McGraw-Hill, New York
5. Bird RB, Stewart WE, Lightfoot EN (1960) Transport phenomena. Wiley, New York

Chapter 7
Application of PDEtoolbox™ to Elasticity Problems

7.1 Basics of Elasticity in Finite Element Application

Consider the free body diagram (Fig. 7.1). Equilibrium of forces yields,

$$\sum F_x = \left(\sigma_x + \frac{\partial \sigma_x}{\partial x}\right) dx\, dy - \sigma_x dx\, dy + \left(\tau_{xy} + \frac{\partial \tau_{xy}}{\partial y}\right) dx\, dy - \tau_{xy} dx\, dy + f_x dx\, dy$$

$$= 0 \tag{7.1}$$

and

$$\sum F_y = \left(\tau_{xy} + \frac{\partial \tau_{xy}}{\partial x}\right) dx\, dy - \tau_{xy} dx\, dy + \left(\sigma_y + \frac{\partial \sigma_y}{\partial x}\right) dx\, dy - \sigma_y dx\, dy + f_y dx\, dy$$

$$= 0 \tag{7.2}$$

where σ_x, σ_y *and* τ_{xy} are stresses acting in the x, y directions and shear stress, respectively. f_x, f_y are body forces/unit area or per unit volume assuming unit thickness.

Thus, simplifying Eqs. 7.1 and 7.2 yields equilibrium equations

$$\frac{\partial \sigma_x}{\partial x} + \frac{\partial \tau_{xy}}{\partial y} + f_x = 0 \tag{7.3}$$

and

$$\frac{\partial \tau_{xy}}{\partial x} + \frac{\partial \sigma_y}{\partial y} + f_y = 0 \tag{7.4}$$

where f_x and f_y are body forces. Note that the surface traction forces come in as natural (Neumann) boundary conditions and diriclet conditions are displacements. Therefore in the weighted integral analysis we can place the surface traction forces as part of the force vector (see Eq. 7.5).

The governing Eqs. 7.3 and 7.4 are then subjected to the weighted integral formulation to derive the finite element equation thus;

O. Oluwole, *Finite Element Modeling for Materials Engineers Using MATLAB*®, DOI: 10.1007/978-0-85729-661-0_7, © Springer-Verlag London Limited 2011

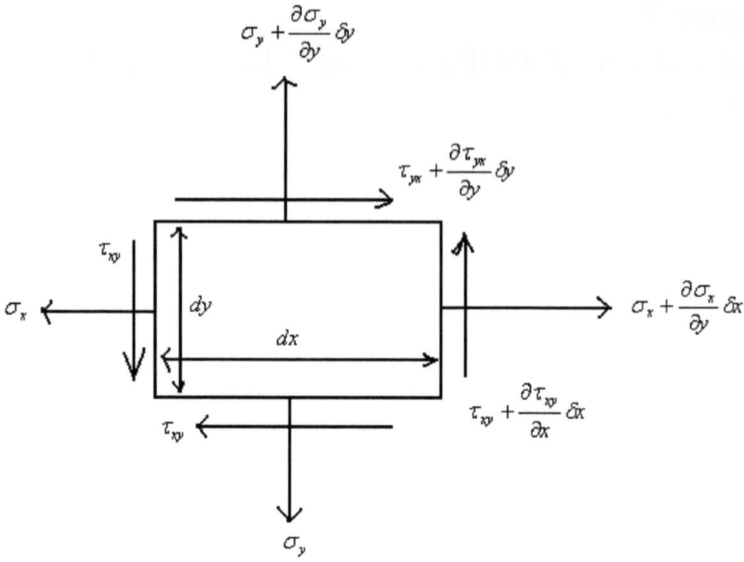

Fig. 7.1 Forces acting on a 2-dimensional infinitesimal elemental body

$$\int_{\Omega} \left\{ \begin{array}{c} w_1 \left(\dfrac{\partial \sigma_x}{\partial x} + \dfrac{\partial \tau_{xy}}{\partial y} \right) \\[2mm] w_2 \left(\dfrac{\partial \tau_{xy}}{\partial x} + \dfrac{\partial \sigma_y}{\partial y} \right) \end{array} \right\} d\Omega + \int_{\Omega} \left\{ \begin{array}{c} w_1 f_x \\ w_2 f_y \end{array} \right\} + \int_{\Gamma_n} \left\{ \begin{array}{c} w_1 \psi_x \\ w_2 \psi_y \end{array} \right\} = 0 \qquad (7.5)$$

where ψ are the traction forces.

Applying integration by parts to the first term gives

$$\int_{\Omega} \left[\begin{array}{ccc} \dfrac{\partial w_1}{\partial x} & 0 & \dfrac{\partial w_1}{\partial y} \\[3mm] 0 & \dfrac{\partial w_2}{\partial y} & \dfrac{\partial w_2}{\partial x} \end{array} \right] \left\{ \begin{array}{c} \sigma_x \\ \sigma_y \\ \tau_{xy} \end{array} \right\} d\Omega + \int_{\Omega} \left\{ \begin{array}{c} w_1 f_x \\ w_2 f_y \end{array} \right\} + \int_{\Gamma_n} \left\{ \begin{array}{c} w_1 \psi_x \\ w_2 \psi_y \end{array} \right\} = 0 \qquad (7.6)$$

The constitutive stress–strain equation is

$$\{\sigma\} = [D]\{\in\} \qquad (7.7)$$

where $\sigma = \left\{ \sigma_x\, \sigma_y\, \tau_{xy} \right\}^T$ and the corresponding strains,

$$\varepsilon = \left\{ \varepsilon_x\, \varepsilon_y\, \gamma_{xy} \right\}^T \qquad (7.8)$$

where $[D]$ is the modulus component expressed as:

$$[D] = \frac{E(1-v)}{(1+v)(1-2v)} \begin{bmatrix} 1 & \dfrac{v}{1-v} & 0 \\[2mm] \dfrac{v}{1-v} & 1 & 0 \\[2mm] 0 & 0 & \dfrac{1-2v}{2(1-v)} \end{bmatrix} \quad \text{for plane strain condition}$$

$$(7.9)$$

and

$$D = \frac{E}{1-v^2} \begin{bmatrix} 1 & v & 0 \\ v & 1 & 0 \\ 0 & 0 & \dfrac{1-v}{2} \end{bmatrix} \quad \text{for plane stress condition} \qquad (7.10)$$

E, v = Elastic modulus and Poisson's ratio, respectively.

The kinematic equations relating strain to displacements are

$$\left\{ \begin{array}{c} \varepsilon_x \\ \varepsilon_y \\ \gamma_{xy} \end{array} \right\} = \left\{ \begin{array}{c} \dfrac{\partial u}{\partial x} \\[2mm] \dfrac{\partial v}{\partial y} \\[2mm] \dfrac{\partial u}{\partial y} + \dfrac{\partial v}{\partial x} \end{array} \right\} = [B]\{d\} \qquad (7.11)$$

where u and v are displacement in the x and y directions, respectively.

In Galerkin finite-element method, weighting function, w is equal to the shape functions, h (see Chap. 4), thus, Eq. 7.11 can be expressed as

$$\left\{ \begin{array}{c} \dfrac{\partial u}{\partial x} \\[2mm] \dfrac{\partial v}{\partial y} \\[2mm] \dfrac{\partial u}{\partial y} + \dfrac{\partial v}{\partial x} \end{array} \right\} = \left\{ \begin{array}{c} \dfrac{\partial w_1}{\partial x} \\[2mm] \dfrac{\partial w_2}{\partial y} \\[2mm] \dfrac{\partial w_1}{\partial y} + \dfrac{\partial w_2}{\partial x} \end{array} \right\} = [B]\{d\} \qquad (7.12)$$

Sustituting Eqs. 7.7–7.11 into 7.6 gives the element equation

$$\int_{\Omega^e} [B]^T [D] [B] \, d\Omega \{d\} + \int_{\Omega^e} \left\{ \begin{array}{c} w_1 f_x \\ w_2 f_y \end{array} \right\} + \int_{\Gamma_{ne}} \left\{ \begin{array}{c} w_1 \psi_x \\ w_2 \psi_y \end{array} \right\} = 0 \qquad (7.13)$$

This is the finite element equation of the form,

$$[k^e]\{d^e\} = \{f^e\} \qquad (7.14)$$

where $[k^e] = \int_{\Omega^e} [B]^T [D] [B] \, d\Omega$ = element stiffness matrix

The shape functions applicable here are C^0 continuous functions since the derivatives are of first order. Linear triangular and bilinear elements are applicable here (see Chap. 3). In the PDEtoolbox, linear triangular element is used.

Linear triangular or rectangular shape functions (Eqs. 3.35 and 3.42) can be substituted into (7.13).

$\{d\} = \{ u_1 \quad v_1 \quad u_2 \quad v_2 \quad u_3 \quad v_3 \}^T$ is the nodal displacement vector
x and y are the directions in which the forces are acting
$E =$ Youngs modulus of elasticity.
$N =$ Poisson ratio.
$\{f^e\} =$ Element force vector.
$\{d^e\} =$ Element displacement vector.

The elemental equations are assembled and solved using the global equation:

$$[k]\backslash\{f\} = \{d\} \tag{7.15}$$

7.2 Using the PDEtoolbox in Modeling Elasticity Problems in Materials Engineering

The general procedure is as elaborated in Chap. 5.

1. Select structural mechanics in the *Options* → *Applications* menu.
2. Draw the domain geometry following directions laid out in Sect. 5.2.3
3. Set the applicable PDE and the initial and boundary conditions. You must be familiar with your governing equation as well as the boundary conditions to successfully model with the PDEtoolbox. In PDEtoolbox application, body forces, f_x and f_y are denoted K_x and K_y and are called volume forces. The Neumann conditions are the surface tractions and spring constants. In the elliptic mode (i.e., for problems of plane strain and plane stress), ρ is taken to be 1.0 because the governing Eq. 7.13 does not involve density. In the eigenvalue mode, however, the value of ρ must be input in the PDE specification form because it is a parameter in the eigenvalue equation (see these texts [1–3] for detailed treatment of the eigenvalue problem).
4. Mesh the domain and refine mesh as applicable using adaptive or uniform meshing. In the adaptive mode, you can use the default options to refine the mesh in areas where better accuracy is needed.
5. Solve the PDE using the *Solve* → *Solve PDE* menu.
6. Explore other solutions using the *Plot* → *Plot Parameters...* menu.

7.3 Applications of PDEtoolbox in Modeling Elasticity Problems

Example 7.1 Stress distribution in metal sheets
Consider the case of a metal sheet 1 m × 1 m under a uniaxial tensile stress of 0.4 MN/m². The plate is clamped on one of its sides. Find the displacements and

Fig. 7.2 *X*-displacement

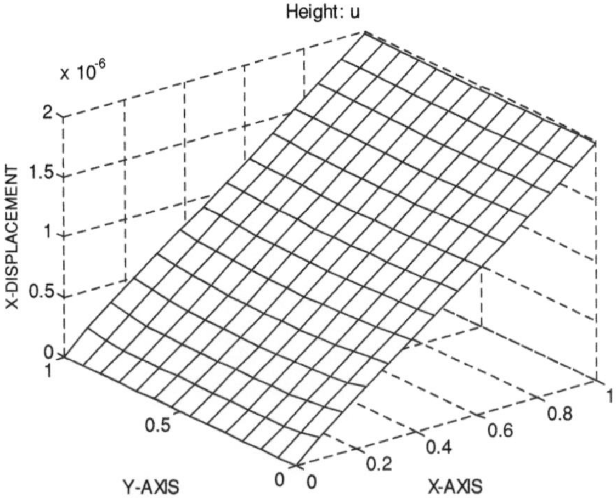

stress and strain distribution in the plate. Also find the Von-Mises displacement. $E = 200E3$ MN/m^2; $v = 0.3$;

Solution

This is an elliptic problem and since it is a sheet, plane stress conditions are operating here.

1. Select *Structural Mechanics–Plane Stress* in the *Options → Applications* menu.
2. Draw the domain using *Options → Grid, Options → Axes Limits and Options → Snap* Select [0 2] for *x*-axis and [0 2] for *y*-axis and draw with a $1 = 1$ scale. Use RECTANGLE to obtain desired domain.
3. Input the boundary conditions. Boundary conditions are the displacements, r_1 and $r_2 = 0.0$ on fixed side. On other free sides select Neumann condition $= 0$ and $g = 0$. On the loaded side select a normal stress of 0.4 (i.e., $q = 0$, $g = 0.4$).
4. Select Elliptic PDE and insert PDE values $E = 200E3$; $v = 0.3$; kx $= 0$; ky $= 0$; $\rho = 1$; there are no body forces.
5. Mesh the domain using meshing procedure elaborated in Sect. 5.6. You do not need to select any parameter on solve parameters. Turn off the adaptive mode.
6. Select the plot parameters. You can use the 3D plot mode. In plotting the deformed mesh, select deformed mesh with desired property (e.g.,) Von-Mises Stress and Deformed mesh displacement in (*u*, *v*) was used in plotting Fig. 7.9.

The solutions are presented in Fig. 7.2, 7.3, 7.4, 7.5, 7.6, 7.7, 7.8, 7.9, 7.10. The codes for this sample problem are presented in *metalsheet.m* file.

Example 7.2 Edge-cracked sheet under plane stress conditions

An edge cracked mild steel specimen 0.30 m × 0.12 m is fixed at one end and stressed 0.4 MN/m^2 at the other end. Find the stress distribution and Von-Mises displacement. $E = 200E3$ MN/m^2; $v = 0.3$;

Fig. 7.3 *Y*-displacement

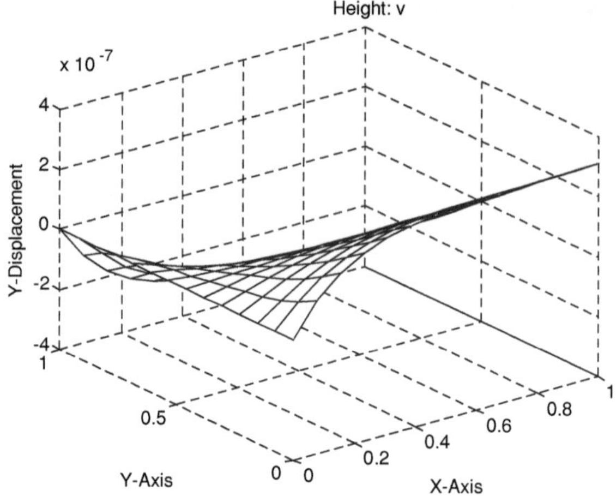

Fig. 7.4 *X*-stress

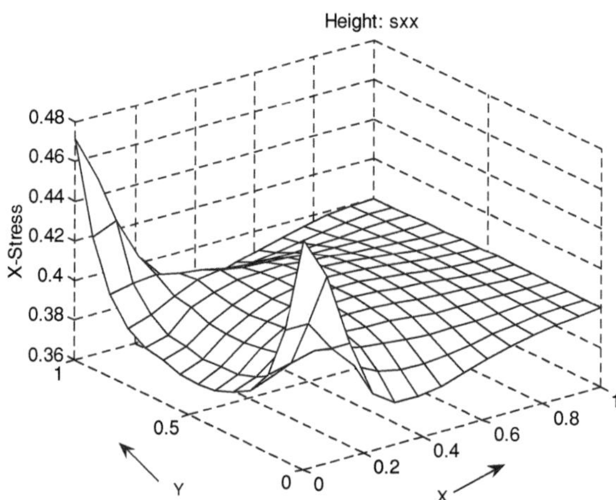

Solution

1. Select *Structural Mechanics–Plane Stress* in the *Options → Applications* menu.
2. Draw the domain using *Options → Axes Limits.* Select [0 0.5] for *x*-axis and [0 0.4] for *y*-axis and draw with a $1 = 1$ scale. Use RECTANGLE-ELLIPSE to obtain desired domain.
3. Input the boundary conditions. Boundary Conditions are the displacements, r_1 and $r_2 = 0.0$, on fixed side. On other free cracked sides select Neumann condition $= 0$ and $g = 0$. On the loaded side select a normal stress of 0.4 (i.e., $q = 0$, $g = 0.4$).
4. Select Elliptic PDE and insert PDE values $E = 200E3$; $v = 0.3$; $kx = 0$; $ky = 0$; $\rho = 1$; there are no body forces.

Fig. 7.5 *Y*-stress

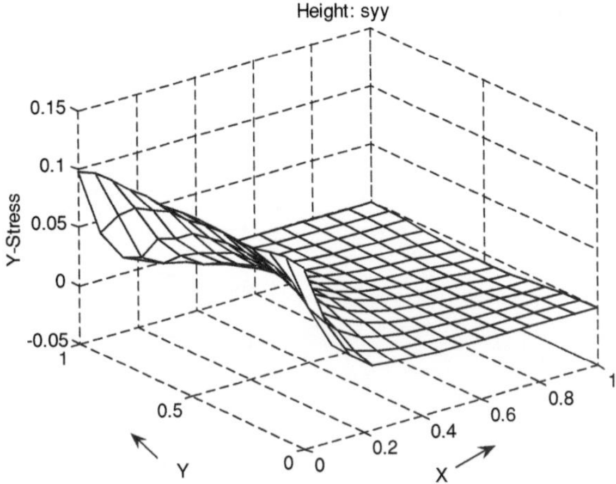

Fig. 7.6 *X*-strain

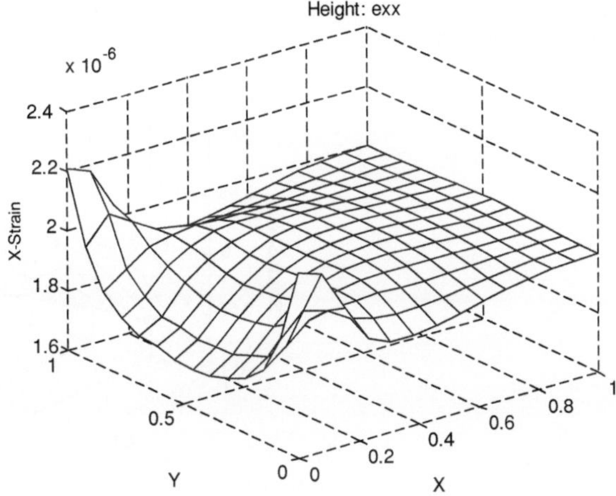

5. Mesh the domain using meshing procedure elaborated in Sect. 5.6. So you do not need to select any parameter on solve parameters. Turn off the adaptive mode.
6. Select the plot parameters. Contour plots have been selected here to fully bring out the effect of stress concentration.

The solution is presented in Figs. 7.11 and 7.12. The M-file for this sample problem is *edgecrackplanestress.m*

Example 7.3 Edge-cracked thick plate under plane strain conditions
An edge cracked mild steel specimen 0.30 m × 0.12 m × 0.12 m is fixed at two opposite ends and stressed 0.4 MN/m^2 at the uncracked end. Find the stress distribution and Von-Mises displacement. $E = 200E3$ MN/m^2; $v = 0.3$

Fig. 7.7 *Y*-strain

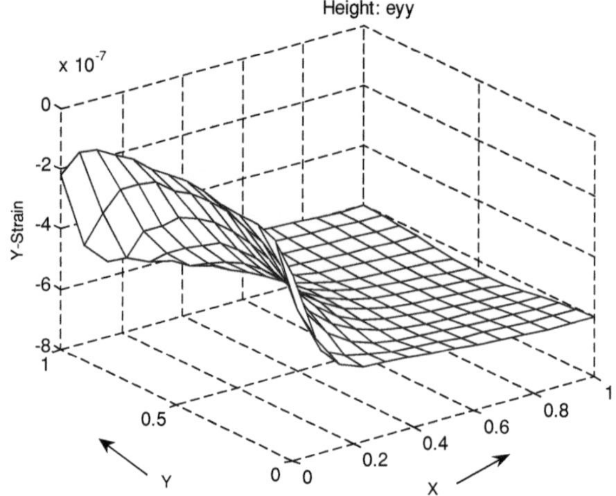

Fig. 7.8 Shear–stress

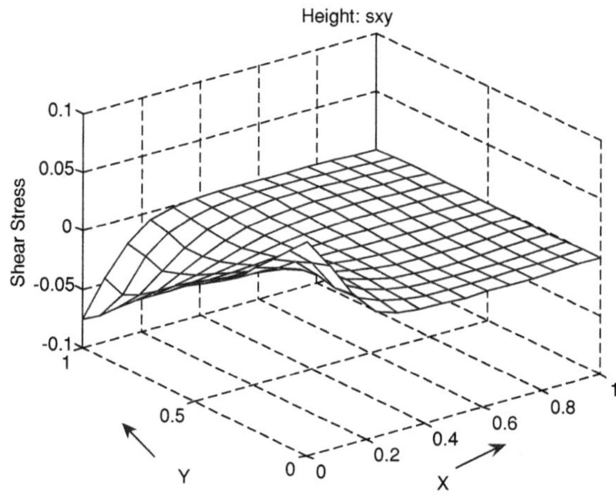

Solution

1. Select *Structural Mechanics–Plane Strain* in the *Options → Applications* menu.
2. Draw the domain using *Options → Axes Limits*. Select [0 0.5] for *x*-axis and [0 0.4] for *y*-axis and draw with a 1 = 1 scale. Use RECTANGLE-ELLIPSE to obtain desired domain.
3. Input the boundary conditions. Boundary conditions are the displacements, r_1 and $r_2 = 0.0$, on fixed sides. On other free side select Neumann condition = 0 and $g = 0$. On the loaded side select a normal stress of 0.4 (i.e., $q = 0$, $g = 0.4$).

Fig. 7.9 Von-Mises stress

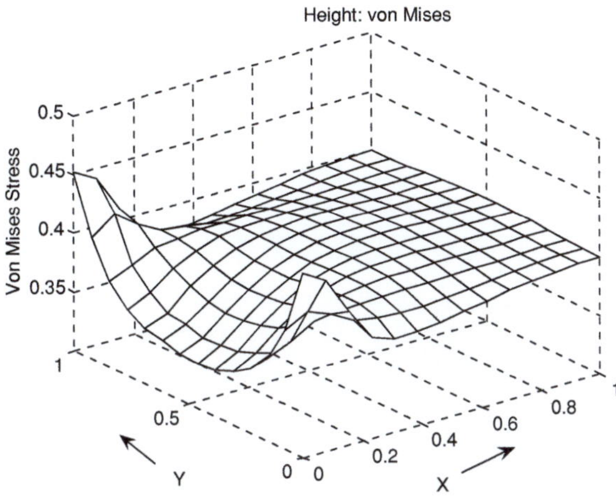

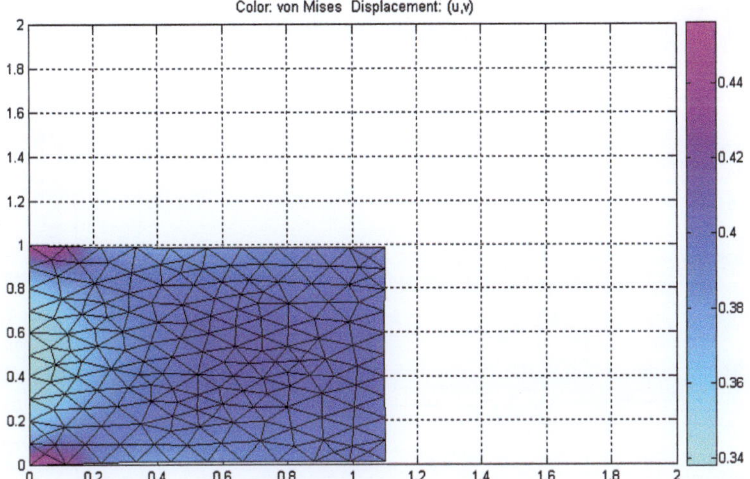

Fig. 7.10 Von-Mises stress and deformation displacement (u, v)

4. Select Elliptic PDE and insert PDE values $E = 200E3$; $v = 0.3$; $kx = 0$, $ky = 0$; $\rho = 1$; there are no body forces.
5. Mesh the domain using meshing procedure elaborated in Sect. 5.6. So you do not need to select any parameter on solve parameters. Turn off the adaptive mode.
6. Select the plot parameters. Contour plots have been selected here to fully bring out the effect of stress concentration.

The solution is presented in Figs. 7.13 and 7.14. The M-file for this sample problem is *Edgecrackplanestrain.m*

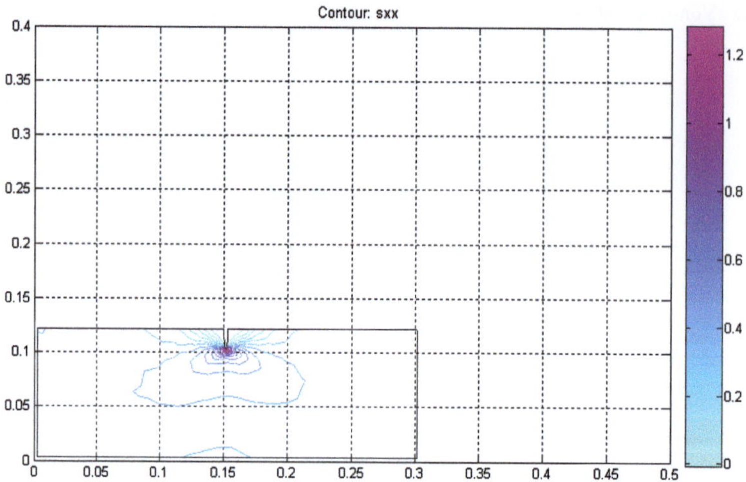

Fig. 7.11 *X*-stress distributions

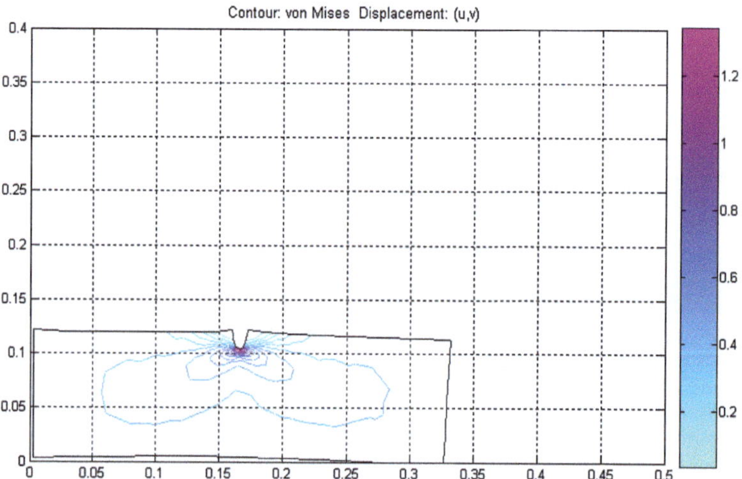

Fig. 7.12 Von-Mises Stress and deformation displacement

Example 7.4 Center cracked thin plate under plane stress conditions
A center-cracked mild steel specimen 0.30 m × 0.12 m is fixed at the left end and stressed 0.4 MN/m^2 at the other end. Find the stress distribution and Von-Mises displacement.

   Use $E = 200E3$; $v = 0.3$; Take body forces to be zero.
   Solution

In center-cracked plates, *Y*-strain is very important because the crack extends in that direction until the crack reaches a value where the plate fails by rapid crack extension.

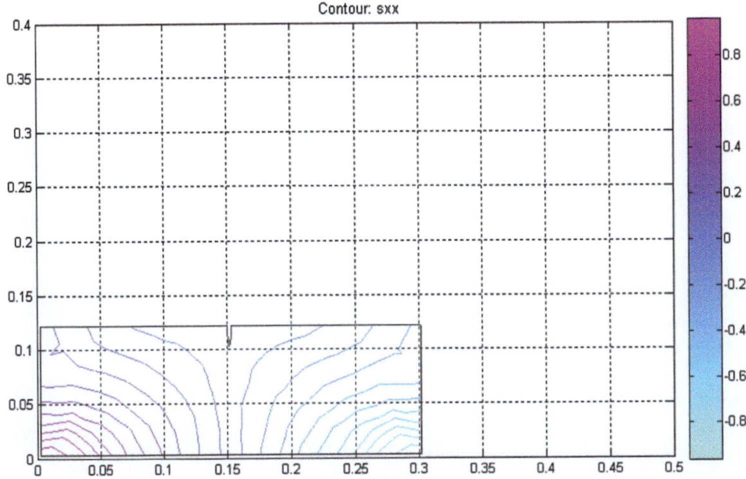

Fig. 7.13 *X*-stress distributions

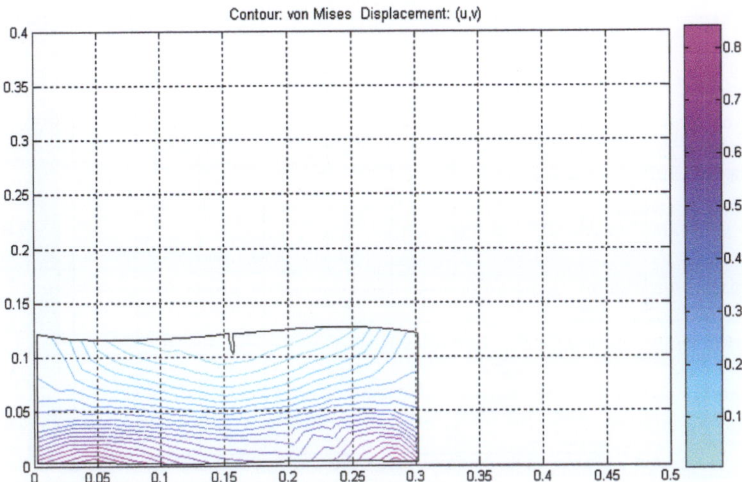

Fig. 7.14 Von-Mises stress and deformation displacement (u, v)

1. Draw the domain using *Options* → *Axes Limits* select axes limits for both x and y axes and draw with a $1 = 1$ scale. Use set formula RECTANGLE-ELLIPSE to obtain desired domain.
2. Input the boundary conditions. Boundary Conditions are the displacements, r_1 and $r_2 = 0.0$, on fixed side. On other free sides select Neumann condition $= 0$ and $g = 0$. On the loaded side select a normal stress of 0.4 (i.e. $q = 0$, $g = 0.4$).

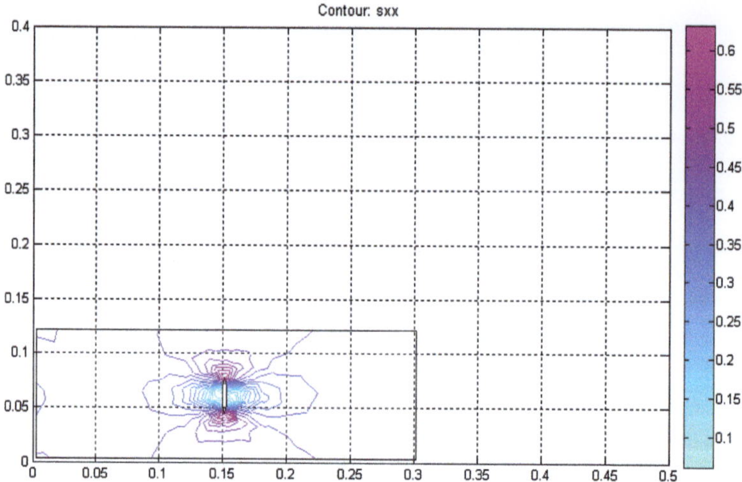

Fig. 7.15 *x*-Stress distribution; notice stress concentration at crack tip

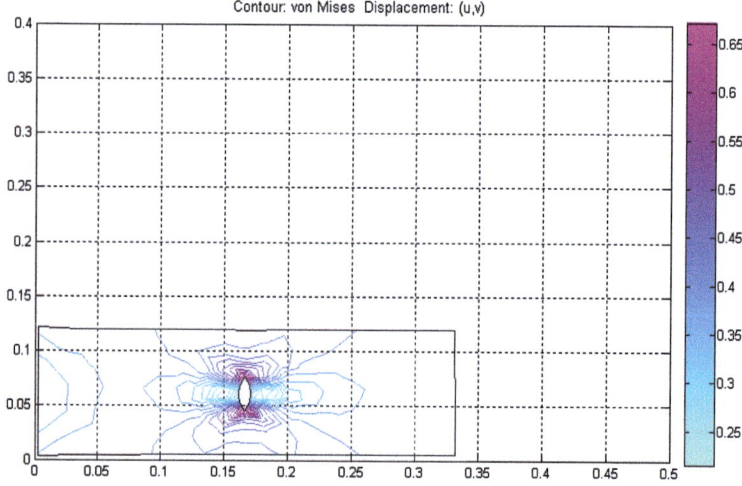

Fig. 7.16 Von-Mises stress and displacement (u, v)

3. Select Elliptic PDE and insert PDE values: $E = 200E3$; $v = 0.3$; kx = 0; ky = 0; $\rho = 1$; there are no body forces.
4. Mesh the domain using meshing procedure elaborated in Sect. 5.6. You do not need to select any parameter on the solve parameters form. Turn off the adaptive mode.
5. Select the plot parameters. In this case *x*-stress.

The solution is presented in Figs. 7.15 and 7.16. The codes for this problem are in *centercrack.m* file.

Example 7.5 Stress effects on ductile iron microstructure (plane strain conditions)
The hole model is traditionally used for modeling nodules in ductile iron [4].

Thus the matrix can be represented by holes using the subtractive ability of PDEtoolGUI. However, we can model the matrix in other ways and see if the model matches the experimental observations.

Question

A ductile iron bar 0.50 m $\times$ 0.15 m $\times$ 0.15 m is fixed at two opposite ends and stressed 0.4MN/m^2 at the one of the free ends. Assuming a twelve nodule microstructure matrix, find the stress and strain distribution in the ductile iron microstructure using (i) the hole model (ii) nodule in a hole model.

Data: Use, $E = 200E3$; $v = 0.3$; kx $= 0$; ky $= 0$ for ductile iron matrix and $E = 10$; $v = 0.15$; kx $= 0$; ky $= 0$; for graphite nodules.

Solution

1. In Draw mode, draw the domain using a scale of $1 = 1$ using *Options* $\rightarrow$ *Axes Limits* select [0 1] for x and [0 0.5] for y-axis, respectively. Use RECTANGLE-ELLIPSE to obtain holes for hole model (Fig. 7.6). For nodule in hole model, use RECTANGLE-ELLIPSES + ELLIPSES (Fig. 7.20). The square is the big square and the subtracted ellipses, the holes. The added small ellipses in the holes are the small nodules. Therefore, we need to use the set formula window below the button icons e.g. *R-E* represents RECTANGLE-ELLIPSE and $R + E$ represents RECTANGLE + ELLIPSE.
2. Input the boundary conditions. Boundary conditions are the displacements, r_1 and $r_2 = 0.0$, on fixed side. On other free sides select Neumann condition $= 0$ and $g = 0$. On the loaded side select a normal stress of 0.4 (i.e., $q = 0$, $g = 0.4$).
3. Select Elliptic PDE and insert PDE values: $E = 200E3$; $v = 0.3$; kx $= 0$; ky $= 0$; $\rho = 1$; there are no body forces.
4. In PDE mode, select Elliptic PDE and insert the PDE values; input values for ductile iron type (i.e., whether ferritic, pearlitic or austenitic steel matrix) in the ductile iron PDE form; input values for graphite in the graphite PDE specification form for nodule in hole model.
5. Mesh the domain using meshing procedure elaborated in Sect. 5.6 (see Fig. 7.8). So you do not need to select any parameter on solve parameters form. Turn off the adaptive mode.
6. Select the plot parameters.

The solution of the hole model is presented in Figs. 7.17, 7.18 and 7.19 and the nodule in hole solution presented in Figs. 7.20 and 7.21. The M-file for this solution is *holeplanestrain.m*.

Example 7.6 Stress effects on ductile Iron microstructure (plane stress conditions)
Question

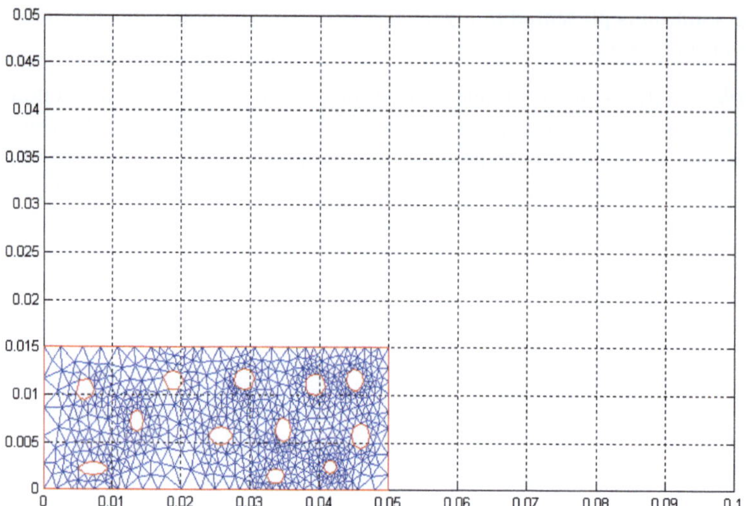

Fig. 7.17　Meshing of the matrix showing ductile iron nodules as holes

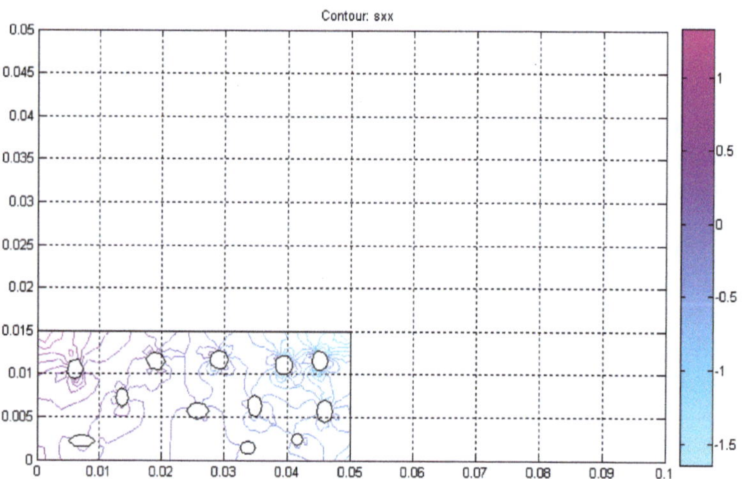

Fig. 7.18　x-stress in a twelve nodule matrix using hole model

Assuming it is possible to obtain a ductile iron sheet 0.50 m × 0.15 m. Find the x-stress, if the sheet is fixed at one end and stressed 0.4 MN/m^2 at the other end. Assuming a twelve nodule microstructure matrix, use (i) the hole model.

Data: Use, $E = 200E3$; $v = 0.3$; kx = 0; ky = 0 for ductile iron matrix and $E = 10$; $v = 0.15$; kx = 0; ky = 0; for graphite nodules.

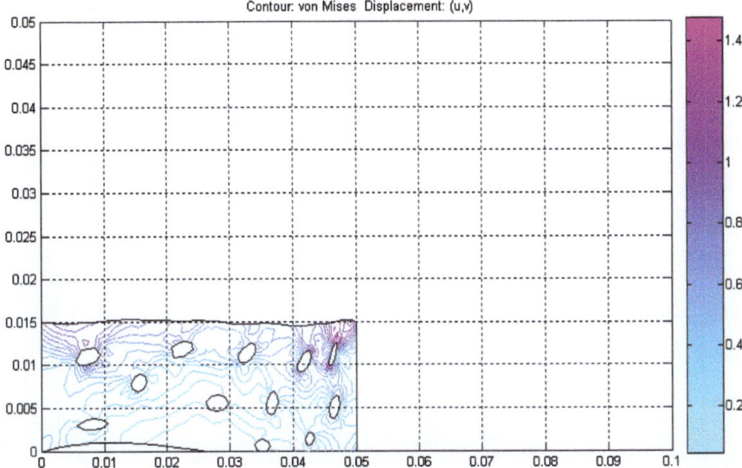

Fig. 7.19 Von-Mises stress and deformation displacement (*u*, *v*)

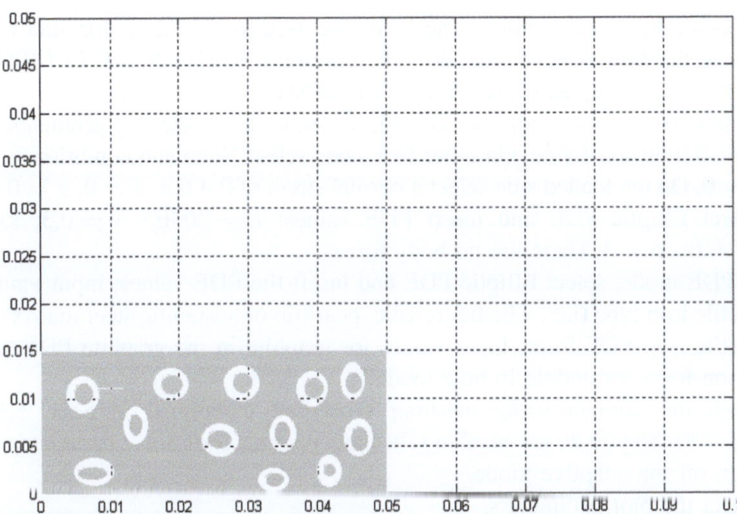

Fig. 7.20 Ductile iron matrix with nodules sitting in holes (nodule in hole model)

Solution

1. In Draw mode, draw the domain using a scale of $1 = 1$ using *Options* $\rightarrow$ *Axes Limits* select [0 1] for *x* and [0 0.5] for *y*-axis, respectively. Use RECTANGLE-ELLIPSE to obtain holes for hole model (Fig. 7.6). For nodule in hole model, use RECTANGLE-ELLIPSES + ELLIPSES (Fig. 7.20). The square is the big

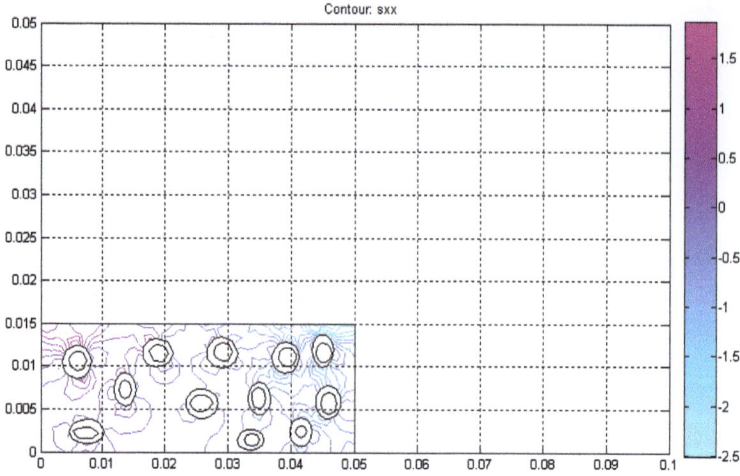

Fig. 7.21 *x*-Stress concentrations at nodule

square and the subtracted ellipses, the holes. The added small ellipses in the holes are the small nodules. Therefore, we need to use the set formula window below the button icons e.g., R − E represents RECTANGLE-ELLIPSE and R + E represents RECTANGLE + ELLIPSE.

2. Input the boundary conditions. Boundary conditions are the displacements, r_1 and $r_2 = 0.0$ on fixed side. On other free sides select Neumann condition $= 0$ and $g = 0$. On the loaded side select a normal stress of 0.4 (i.e. $q = 0$, $g = 0.4$).
3. Select Elliptic PDE and insert PDE values: $E = 200E3$; $v = 0.3$; kx $= 0$; ky $= 0$; $\rho = 1$ There are no body forces.
4. In PDE mode, select Elliptic PDE and insert the PDE values; input values for ductile iron type (i.e., whether ferritic, pearlitic or austenitic steel matrix) in the ductile iron PDE form; input values for graphite in the graphite PDE specification form for nodule in hole model.
5. Mesh the domain using meshing procedure elaborated in Sect. 5.6 (see Fig. 7.8). So you do not need to select any parameter on solve parameters form. Turn off the adaptive mode.
6. Select the plot parameters.

The M-files for this solution is *holeplanestress.m.*

Example 7.7 stress concentrations in hypoeutectoid steels under plane strain conditions (example mild steel)

In this example effect of different phase constituents on overall stress distribution in the metal matrix is examined. This example can also be used in studying the effect of distributed phases in materials through ageing. Thus, new materials can be designed based on effective models developed through the finite element modeling [5].

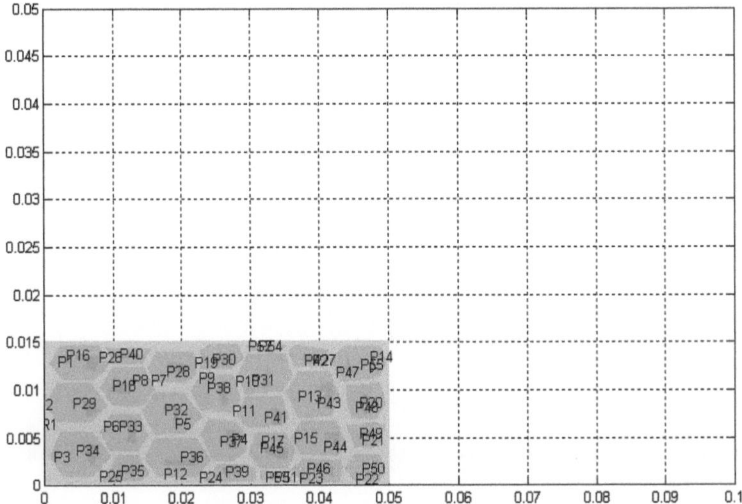

Fig. 7.22 Mild steel microstructure modeling using PDEtool showing pearlite and ferrite phases and grain boundaries

A mild steel bar 0.50 m by 0.15 m $\times$ 0.15 is constrained on two opposite sides and stressed 0.4 MN/m^2 at one of the free sides. Find the deformation displacement and x-stress.

Data: Use, $E = 200E3$; $v = 0.3$; kx $= 0$; ky $= 0$ for ferrite phase and grain boundaries; $E = 250E3$; $v = 0.3$; kx $= 0$; ky $= 0$ for pearlite phase. The M-file for this program is in *microstructureplanestrain.m*.

Solution

1. Draw the domain using a scale of 1:1 using *Options* → *Axes Limits* select [0 1] for x and [0 0.5] for y-axis, respectively. Use SQUARE + POLY-GONS + POLYGONS to obtain microstructure (grains, Ferrite and pearlite phases and grain boundaries (Fig. 7.20)). The square is the domain and the added polygons, the grains. The added small polygons or ellipse in the grains are the small pearlite phases.
2. Input the boundary conditions. Boundary conditions are the displacements, r_1 and $r_2 = 0.0$, on fixed sides. On other free side select Neumann condition $= 0$ and $g = 0$. On the loaded side select a normal stress of 0.4 (i.e., $q = 0$, $g = 0.4$).
3. Select Elliptic PDE and insert PDE values: $E = 200E3$; $v = 0.3$; kx $= 0$; ky $= 0$; $\rho = 1$; there are no body forces. Select Elliptic PDE and insert the PDE values. Input values for ferritic steel in the ferrite phase; input values for pearltic steel in the pearlite phase; $E = 200E3$; $v = 0.3$; kx $= 0$; ky $= 0$ for ferrite phase and grain boundaries; $E = 250E3$; $v = 0.3$; kx $= 0$; ky $= 0$ for pearlite phase.
4. Mesh the domain using uniform meshing procedure elaborated in Sect. 5.6. You do not need to select any parameter on solve parameters. Turn off the adaptive mode.

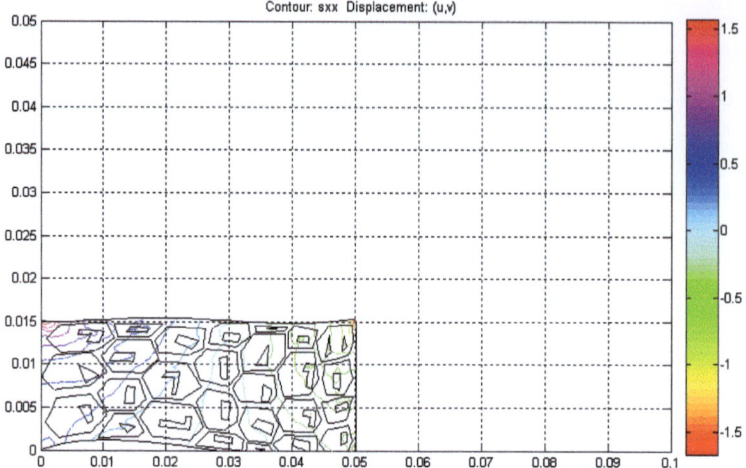

Fig. 7.23 *x*-Stress distribution and deformation displacement pattern under plane strain

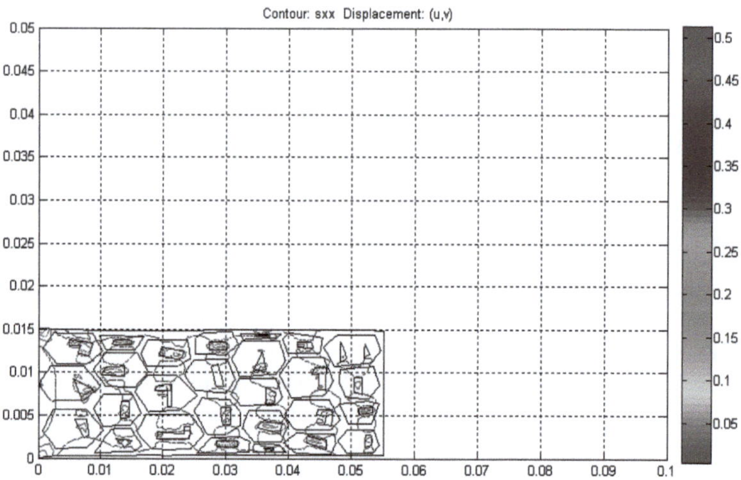

Fig. 7.24 *x*-Stress and deformation displacement under plane stress

5. Select the plot parameters. In this example let us use HSV color to bring out the different color dilineations in the metal matrix. The solutions are presented in Figs. 7.22 and 7.23.

Example 7.8 stress concentrations in hypoeutectoid steels under plane stress conditions (example mild steel)
A mild steel sheet 0.50 m × 0.15 m fixed at one end and stressed 0.4 MN/m^2 at

the other end. Find the deformation displacement and the x-stress. Data: Use, $E = 200E3$; $v = 0.3$; kx $= 0$; ky $= 0$ for ferrite phase and grain boundaries; $E = 250E3$; $v = 0.3$; kx $= 0$; ky $= 0$ for pearlite phase.

Solution

1. Draw the domain using a scale of 1:1 using *Options* → *Axes Limits* select [0 1] for x and [0 0.5] for y-axis, respectively. Use SQUARE + POLYGONS + POLYGONS to obtain microstructure (grains, Ferrite and pearlite phases and grain boundaries (Fig. 7.20). The square is the domain and the added polygons, the grains. The added small polygons or ellipse in the grains are the small pearlite phases.
2. Input the boundary conditions. Boundary conditions are the displacements, r_1 and $r_2 = 0.0$, on fixed side. On other free sides select Neumann condition $= 0$ and $g = 0$. On the loaded side select a normal stress of 0.4 (i.e., $q = 0$, $g = 0.4$).
3. Select Elliptic PDE and insert PDE values: $E = 200E3$; $v = 0.3$; kx $= 0$; ky $= 0$; $\rho = 1$; there are no body forces. Select Elliptic PDE and insert the PDE values. Input values for ferritic steel in the ferrite phase; input values for pearlitic steel in the pearlite phase; $E = 200E3$; $v = 0.3$; kx $= 0$; ky $= 0$ for ferrite phase and grain boundaries; $E = 250E3$; $v = 0.3$; kx $= 0$; ky $= 0$ for pearlite phase.
4. Mesh the domain using uniform meshing procedure elaborated in Sect. 5.6 (see Fig. 7.26). You do not need to select any parameter on solve parameters. Turn off the adaptive mode.
5. Select the plot parameters. In this example let us use HSV color to bring out the different color dilineations in the metal matrix.

The solutions are presented in Fig. 7.24. The M-file is presented in *microstructureplanestress.m* file.

7.4 Exercises

6.1 Obtain x-displacement deformed mesh and x-strain deformed mesh for Example 7.1 using *metal sheet m*-file.

6.2 Using *edgecrackplanestress* m-file obtain the shear stress and shear strain. Also obtain the deformed mesh for displacement, and x and y strains. Extract the stress value at the crack tip.

6.3 Find the x-Stress, y-strain and shear stress distributions with a displacement of 0.01 at the sides of Example 7.3. Use the *m*-file *centercrack*.

6.4 Obtain shear stress for both hole model and nodule in hole model and compare results. Use the *m*-files *holeplanestrain* and *noduleplanestrain* provided in the CD.

6.5 Repeat *holeplanestrain* with an 8-nodule matrix. What do you observe? Study the x-stress, x-strain and the shear stress and shear strain distributions in the matrix.

6.6 Repeat nodule in hole model with an 8-nodule matrix using *noduleplanestrain* *m*-file and compare with results of hole model for a 12-nodule matrix.

6.7 Repeat the *hole model*, this time instead of holes, use actual graphite nodules by changing the subtractive signs in the *Set Formula* window to plus. Input values for graphites in the PDE specification form.

6.8 Work out Example 7.6 for nodule in hole model subjected to plane stress.

6.9 (i) Find the shear stress and the Von-Mises stress for Example 7.8. Use the *microstructureplanestress m*-file

(ii) Using *microstructureplanestress m*-file, add more pearlite into the matrix structure and see this affects stress distribution and possible fracture morphology.

References

1. Timoshenko SP, Goodier JN (1970) Theory of elasticity, 3rd edn. McGraw-Hill, Singapore
2. Reddy JN (2006) An introduction to the finite element method, 3rd edn. McGraw-Hill, New York
3. Stasa FL (1985) Applied finite element analysis for engineers. CBS Publishing Japan Ltd, New York
4. Pundale SH, Rogers RJ, Nadkarni GR (1998) Finite element modeling of elastic modulus in ductile irons: effect of graphite morphology. Trans Am Foundrym Soc 98(102):98–105
5. Oluwole OO (2009) Stress loops effect in ductile failure of mild steel. J Miner Mater Charact Eng 8(4):293–302

Appendix: Sample M-codes for Exercises in the Text

METAL.m
```
function pdemodel
[pde_fig,ax]=pdeinit;
pdetool('appl_cb',9);
pdetool('snapon','on');
set(ax,'DataAspectRatio',[1 0.050000000000000003 1]);
set(ax,'PlotBoxAspectRatio',[30 20 1]);
set(ax,'XLim',[0 60]);
set(ax,'YLim',[0 2]);
set(ax,'XTickMode','auto');
set(ax,'YTickMode','auto');
pdetool('gridon','on');
% Geometry description:
pderect([0 50 1 0],'METAL');
set(findobj(get(pde_fig,'Children'),'Tag','PDEEval'),'String','METAL')
% Boundary conditions:
pdetool('changemode',0)
pdesetbd(4,...
'dir',...
1,...
'1',...
'400')
pdesetbd(3,...
'neu',...
1,...
'0',...
'0')
pdesetbd(2,...
'dir',...
1,...
'1',...
```

O. Oluwole, *Finite Element Modeling for Materials Engineers Using MATLAB*®,
DOI: 10.1007/978-0-85729-661-0, © Springer-Verlag London Limited 2011

```
'25')
pdesetbd(1,...
'neu',...
1,...
'0',...
'0')
% Mesh generation:
setappdata(pde_fig,'Hgrad',1.3);
setappdata(pde_fig,'refinemethod','regular');
pdetool('initmesh')
pdetool('refine')
pdetool('refine')
% PDE coefficients:
pdeseteq(1,...
'73.0',...
'0.0',...
'(0.0)+(0.0).*(0.0)',...
'(1.0).*(1.0)',...
'0:10',...
'0.0',...
'0.0',...
'[0 100]')
setappdata(pde_fig,'currparam',...
['1.0 ';...
'1.0 ';...
'73.0';...
'0.0 ';...
'0 ';...
'0.0 '])
% Solve parameters:
setappdata(pde_fig,'solveparam',...
str2mat('0','1000','10','pdeadworst',...
'0.5','longest','0','1E-4','','fixed','Inf'))
% Plotflags and user data strings:
setappdata(pde_fig,'plotflags',[1 1 1 1 1 1 1 1 1 0 0 0 1 1 0 0 0 0 1]);
setappdata(pde_fig,'colstring',);
setappdata(pde_fig,'arrowstring',);
setappdata(pde_fig,'deformstring',);
setappdata(pde_fig,'heightstring','');
% Solve PDE:
pdetool('solve')
```

DIRICLET.m
```
function pdemodel
[pde_fig,ax]=pdeinit;
```

```
pdetool('appl_cb',9);
pdetool('snapon','on');
set(ax,'DataAspectRatio',[1 1.5 1]);
set(ax,'PlotBoxAspectRatio',[1 0.66666666666666663 1]);
set(ax,'XLim',[0 2]);
set(ax,'YLim',[0 2]);
set(ax,'XTickMode','auto');
set(ax,'YTickMode','auto');
pdetool('gridon','on');
% Geometry description:
pderect([0 1 1 0],'SQ1');
set(findobj(get(pde_fig,'Children'),'Tag','PDEEval'),'String','SQ1')
% Boundary conditions:
pdetool('changemode',0)
pdesetbd(4,...
'dir',...
1,...
'1',...
'100')
pdesetbd(3,...
'dir',...
1,...
'1',...
'500')
pdesetbd(2,...
'dir',...
1,...
'1',...
'25')
pdesetbd(1,...
'dir',...
1,..
'1',...
'250')
% Mesh generation:
setappdata(pde_fig,'Hgrad',1.3);
setappdata(pde_fig,'refinemethod','regular');
pdetool('initmesh')
% PDE coefficients:
pdeseteq(2,...
'73.4',...
'28.39',...
'(0)+(28.39).*(25.0)',...
'(7680.0).*(1300)',...
'0:3000',...
```

```
‘25.0’,...
‘0.0’,...
‘[0 100]’)
setappdata(pde_fig,’currparam’,...
[‘7680.0’;...
‘1300 ’;...
‘73.4’;...
‘0 ’’;...
‘28.39 ’;...
‘25.0 ’])
% Solve parameters:
setappdata(pde_fig,‘solveparam’,...
str2mat(‘0’,‘1000’,‘10’,‘pdeadworst’,...
‘0.5’,‘longest’,‘0’,‘1E-4’,’’,‘fixed’,‘Inf’))
% Plotflags and user data strings:
setappdata(pde_fig,‘plotflags’,[1 1 1 1 1 1 1 1 0 0 0 3001 0 1 0 0 0 1]);
setappdata(pde_fig,‘colstring’,’’);
setappdata(pde_fig,‘arrowstring’,’’);
setappdata(pde_fig,‘deformstring’,’’);
setappdata(pde_fig,‘heightstring’,’’);
% Solve PDE:
pdetool(‘solve’)
```

TRANSIENTTUBULAR.m

```
function pdemodel
[pde_fig,ax]=pdeinit;
pdetool(‘appl_cb’,9);
set(ax,‘DataAspectRatio’,[1 1 1]);
set(ax,‘PlotBoxAspectRatio’,[1.5 1 1]);
set(ax,‘XLim’,[0 3]);
set(ax,‘YLim’,[0 2]);
set(ax,‘XTickMode’,‘auto’);
set(ax,‘YTickMode’,‘auto’);
% Geometry description:
pdeellip(0.55725190839694694,0.52328244274809221,
    0.54045801526717518,0.49923664122137401,...
0,‘E1’);
pdeellip(0.55267175572519056,0.52328244274809221,
    0.50839694656488543,0.47633587786259546,...
0,‘E2’);
pdecirc(0.55725190839694649,0.53702290076335935,
    0.22900763358778642,‘C1’);
pdeellip(0.55725190839694649,0.53702290076335935,
    0.28396946564885495,0.27938931297709924,...
0,‘E3’);
```

```
set(findobj(get(pde_fig,'Children'),'Tag','PDEEval'),
  'String','E1+E2+C1+E3')
% Boundary conditions:
pdetool('changemode',0)
pdesetbd(8,...
'dir',...
1,...
'1',...
'30')
pdesetbd(7,...
'dir',...
1,...
'1',...
'30')
pdesetbd(6,...
'dir',...
1,...
'1',...
'30')
pdesetbd(5,...
'dir',...
1,...
'1',...
'30')
% Mesh generation:
setappdata(pde_fig,'Hgrad',1.3);
setappdata(pde_fig,'refinemethod','regular');
pdetool('initmesh')
% PDE coefficients:
pdeseteq(2,...
'0.15!0.6!0.04!0',...
'28.39!0!0!28.39',...
'(715.0)+(28.39).*(30+t)!(0)+(0).*(0)!(0)+(0).*(0)!(0)+(28.39).*(30+t)',...
'(1.0).*(1.0)!(1.0).*(1.0)!(1.0).*(1.0)!(1.0).*(1.0)',...
'0.1000',...
'30',...
'0.0',...
'[0 100]')
setappdata(pde_fig,'currparam',...
['1.0!1.0!1.0!1.0';...
'1.0!1.0!1.0!1.0';...
'.15!0.6!0.04!0';...
'715.0!0!0!0 ';...
'28.39!0!0!28.39';...
'30+t!0!0!30+t '])
```

```
% Solve parameters:
setappdata(pde_fig,'solveparam',...
str2mat('0','1988','10','pdeadworst',...
'0.5','longest','0','1E-4','','fixed','Inf'))
% Plotflags and user data strings:
setappdata(pde_fig,'plotflags',[1 1 1 1 1 1 1 1 0 0 0 1001 0 1 0 0 0 1]);
setappdata(pde_fig,'colstring','');
setappdata(pde_fig,'arrowstring','');
setappdata(pde_fig,'deformstrin','');
setappdata(pde_fig,'heightstring','');
% Solve PDE:
pdetool('solve')
```

STEADYSTATETUBULAR.m

```
function pdemodel
[pde_fig,ax]=pdeinit;
pdetool('appl_cb',9);
set(ax,'DataAspectRatio',[1 1 1]);
set(ax,'PlotBoxAspectRatio',[1.5 1 1]);
set(ax,'XLim',[0 3]);
set(ax,'YLim',[0 2]);
set(ax,'XTickMode','auto');
set(ax,'YTickMode','auto');
% Geometry description:
pdeellip(0.55725190839694694,0.52328244274809221,
  0.54045801526717518,0.49923664122137401,...
0,'E1');
pdeellip(0.55267175572519056,0.52328244274809221,
  0.50839694656488543,0.47633587786259546,...
0,'E2');
pdecirc(0.55725190839694649,0.53702290076335935,
  0.22900763358778642,'C1');
pdeellip(0.55725190839694649,0.53702290076335935,
  0.28396946564885495,0.27938931297709924,...
0,'E3');
set(findobj(get(pde_fig,'Children'),'Tag','PDEEval'),
  'String','E1+E2+C1+E3')
% Boundary conditions:
pdetool('changemode',0)
pdesetbd(8,...
'dir',...
1,...
'1',...
'30')
pdesetbd(7,...
```

```
'dir',...
1,...
'1',...
'30')
pdesetbd(6,...
'dir',...
1,...
'1',...
'30')
pdesetbd(5,...
'dir',...
1,...
'1',...
'30')
% Mesh generation:
setappdata(pde_fig,'Hgrad',1.3);
setappdata(pde_fig,'refinemethod','regular');
pdetool('initmesh')
% PDE coefficients:
pdeseteq(1,...
'0.15!0.6!0.04!0',...
'28.39!0!0!28.39',...
'(715.0)+(28.39).*(1200)!(0)+(0).*(0)!(0)+(0).*(0)!(0)+(28.39).*(1200)',...
'(1.0).*(1.0)!(1.0).*(1.0)!(1.0).*(1.0)!(1.0).*(1.0)',...
'0:1000',...
'30',...
'0.0',...
'[0 100]')
setappdata(pde_fig,'currparam',...
['1.0!1.0!1.0!1.0';...
'1.0!1.0!1.0!1.0';...
'0.15!0.6!0.04!0';...
'715.0!0!0!0 ';...
'28.39!0!0!28.39';...
'1200!0!0!1200 '])
% Solve parameters:
setappdata(pde_fig,'solveparam',...
str2mat('0','1988','10','pdeadworst',...
'0.5','longest','0','1E-4','','fixed','Inf'))
% Plotflags and user data strings:
setappdata(pde_fig,'plotflags',[1 1 1 1 1 1 1 1 0 0 0 1 0 1 0 0 0 1]);
setappdata(pde_fig,'colstring','');
setappdata(pde_fig,'arrowstring','');
setappdata(pde_fig,'deformstring','');
setappdata(pde_fig,'heightstring','');
```

```matlab
% Solve PDE:
pdetool('solve')
```

METALSHEET.m

```matlab
function pdemodel
[pde_fig,ax]=pdeinit;
pdetool('appl_cb',3);
pdetool('snapon','on');
set(ax,'DataAspectRatio',[1 1.5 1]);
set(ax,'PlotBoxAspectRatio',[1 0.66666666666666663 1]);
set(ax,'XLim',[0 2]);
set(ax,'YLim',[0 2]);
set(ax,'XTickMode','auto');
set(ax,'YTickMode','auto');
pdetool('gridon','on');
% Geometry description:
pderect([0 1 1 0],'SQ1');
set(findobj(get(pde_fig,'Children'),'Tag','PDEEval'),'String','SQ1')
% Boundary conditions:
pdetool('changemode',0)
pdesetbd(4,...
'dir',...
2,...
str2mat('1','0','0','1'),...
str2mat('0','0'))
pdesetbd(3,...
'neu',...
2,...
str2mat('0','0','0','0'),...
str2mat('0','0'))
pdesetbd(2,...
'neu',...
2,...
str2mat('0','0','0','0'),...
str2mat('0.4','0'))
pdesetbd(1,...
'neu',...
2,...
str2mat('0','0','0','0'),...
str2mat('0','0'))
% Mesh generation:
setappdata(pde_fig,'Hgrad',1.3);
setappdata(pde_fig,'refinemethod','regular');
pdetool('initmesh')
% PDE coefficients:
```

```
pdeseteq(1,...
str2mat('2*((200E3)./(2*(1+(0.3))))+(2*((200E3)./(2*(1+(0.3))))).*(0.3)./(1-
(0.3)))','0','(200E3)./(2*(1+(0.3)))','0','(200E3)./(2*(1+(0.3)))','2*((200E3)./
  (2*(1+(0.3)))).*(0.3)./(1-
(0.3)))','0','(200E3)./(2*(1+(0.3)))','0','2*((200E3)./
  (2*(1+(0.3))))+(2*((200E3)./(2*(1+(0.3))))).*(0.3)./(1-
(0.3)))'),...
str2mat('0.0','0.0','0.0','0.0'),...
str2mat('0.0','0.0'),...
str2mat('1.0','0','0','1.0'),...
'0:10',...
'0.0',...
'0.0',...
'[0 100]')
setappdata(pde_fig,'currparam',...
['200E3';...
'0.3';...
'0.0';...
'0.0';...
'1.0'])
% Solve parameters:
setappdata(pde_fig,'solveparam',...
str2mat('0','1000','10','pdeadworst',...
'0.5','longest','0','1e-4','','fixed'','inf'))
% Plotflags and user data strings:
setappdata(pde_fig,'plotflags',[18 1 18 1 1 1 1 1 0 1 0 1 1 0 0 0 1 1]);
setappdata(pde_fig,'colstring','');
setappdata(pde_fig,'arrowstring','');
setappdata(pde_fig,'deformstring','');
setappdata(pde_fig,'heightstring','');
% Solve PDE:
pdetool('solve')
```

CENTERCRACK.m

```
function pdemodel
[pde_fig,ax]=pdeinit;
pdetool('appl_cb',3);
set(ax,'DataAspectRati',[1 1.2000000000000002 1]);
set(ax,'PlotBoxAspectRatio',[1.25 0.833333333333333326 5]);
set(ax,'XLim',[0 0.5]);
set(ax,'YLim',[0 0.40000000000000002]);
set(ax,'XTickMode','auto');
set(ax,'YTickMode','auto');
pdetool('gridon','on');
% Geometry description:
```

```
pdeellip(-
0.2587786259541843,0.41450381679389359,0.011450381679389166,
   0.066412213740458026,...
0,'E1');
pderect([0.0026717557251908775 0.30190839694656491
   0.12122137404580148
0.003053435114503789],'R1');
pdeellip(0.15152671755725192,0.060763358778625931,
   0.001526717557251922,0.0164885496183206
22,...
0,'E2');
set(findobj(get(pde_fig,'Children'),'Tag','PDEEval'),'String','R1-E2')
% Boundary conditions:
pdetool('changemode',0)
pdesetbd(8,...
'neu',...
2,...
str2mat('0','0','0','0'),...
str2mat('0','0'))
pdesetbd(7,...
'neu',...
2,...
str2mat('0','0','0','0'),...
str2mat('0','0'))
pdesetbd(6,...
'neu',...
2,...
str2mat('0','0','0','0'),...
str2mat('0','0'))
pdesetbd(5,...
'neu',...
2,...
str2mat('0','0','0','0'),...
str2mat('0','0'))
pdesetbd(4,...
'neu',...
2,...
str2mat('0','0','0','0'),...
str2mat('0','0'))
pdesetbd(3,...
'dir',...
2,...
str2mat('1','0','0','1'),...
str2mat('0','0'))
pdesetbd(2,...
```

```
vneu',...
2,...
str2mat('0','0','0','0'),...
str2mat('0','0'))
pdesetbd(1,...
'neu',...
2,...
str2mat('0','0','0','0'),...
str2mat('0.4','0'))
% Mesh generation:
setappdata(pde_fig,'Hgrad',1.3);
setappdata(pde_fig,'refinemethod','regular');
pdetool('initmesh')
% PDE coefficients:
pdeseteq(1,...
str2mat('2*((200E3)./(2*(1+(0.3))))+(2*((200E3)./(2*(1+(0.3))))).*(0.3)./(1-
(0.3)))','0','200E3)./(2*(1+(0.3)))','0','(200E3)./(2*(1+(0.3)))',v2*((200E3)./
   (2*(1+(0.3)))).*(0.3)./(1-
(0.3)))','0','(200E3)./(2*(1+(0.3)))','0','2*((200E3)./
   (2*(1+(0.3))))+(2*((200E3)./(2*(1+(0.3))))).*(0.3)./(1
(0.3)))'),..
str2mat('0.0','0.0','0.0','0.0'),...
str2mat('0.0','0.0'),...
str2mat('1.0','0','0','1.0'),...
'0:10',...
'0.0',..
'0.0',...
'[0 100]')
setappdata(pde_fig,'currparam',..
['200E3'';...
'0.3 ';...
'0.0 ';...
'0.0 ';...
'1.0 '])
% Solve parameters
setappdata(pde_fig,'solveparam',...
str2mat('0','1000','10','pdeadworst',...
'0.5','longest','0','1e-4','','fixed','inf'))
% Plotflags and user data strings:
setappdata(pde_fig,'plotflags',[18 1 1 1 1 1 1 1 0 1 0 1 0 1 0 0 1 1]);
setappdata(pde_fig,'colstring','');
setappdata(pde_fig,'arrowstring','');
setappdata(pde_fig,'deformstring','');
setappdata(pde_fig,'heightstring','');
% Solve PDE:
```

```
pdetool('solve')
```

EDGECRACKPLANESTRESS.m

```
function pdemodel
[pde_fig,ax]=pdeinit;
pdetool('appl_cb',3);
set(ax,'DataAspectRatio',[1 1.2000000000000002 1]);
set(ax,'PlotBoxAspectRatio',[1.25 0.83333333333333326 5]);
set(ax,'XLim',[0 0.5]);
set(ax,'YLim',[0 0.40000000000000002]);
set(ax,'XTickMode','auto');
set(ax,'YTickMode','auto');
pdetool('gridon','on');
% Geometry description:
pdeellip(-
0.25877862595419843,0.41450381679389359,
   0.011450381679389166,0.066412213740458026,..
0,'E1');
pderect([0.0026717557251908775 0.30190839694656491
   0.12122137404580148
0.0030534351145037889],'R1');
pdeellip(0.15152671755725192,0.1193893129770992,
   0.0015267175572512922,0.016488549618320622,...
0,'E2');
set(findobj(get(pde_fig,'Children'),'Tag','PDEEval'),'String','R1-E2')
% Boundary conditions:
pdetool('changemode',0)
pdesetbd(8,...
'neu',...
2,...
str2mat('0','0','0','0'),...
str2mat('0','0'))
pdesetbd(7,...
'neu',...
2,...
str2mat('0','0','0','0'),...
str2mat('0','0'))
pdesetbd(6,...
'neu',...
2,...
str2mat('0','0','0','0'),...
str2mat('0','0'))
pdesetbd(5,...
'neu',...
2,...
```

```
str2mat('0','0','0','0'),...
str2mat('0.4','0'))
pdesetbd(4,...
'neu',...
2,...
str2mat('0','0','0','0'),...
str2mat('0','0'))
pdesetbd(3,...
'neu',...
2,...
str2mat('0','0','0','0'),...
str2mat('0','0'))
pdesetbd(2,...
'dir',...
2,...
str2mat('1','0','0','1'),...
str2mat('0','0'))
pdesetbd(1,...
'neu',...
2,...
str2mat('0','0','0','0'),...
str2mat('0','0'))
% Mesh generation:
setappdata(pde_fig,'Hgrad',1.3);
setappdata(pde_fig,'refinemethod','regular');
pdetool('initmesh')
% PDE coefficients:
pdeseteq(1,...
str2mat('2*((200E3)./(2*(1+(0.3))))+(2*((200E3)./(2*(1+(0.3))))).*(0.3)./(1-
(0.3)))','0','(200E3)./(2*(1+(0.3)))','0','(200E3)./(2*(1+(0.3)))','2*((200E3)./
   (2*(1+(0.3)))).*(0.3)./(1-
(0.3)))','0','(200E3)./(2*(1+(0.3)))','0','2*((200E3)./
   (2*(1+(0.3))))+(2*((200E3)./(2*(1+(0.3))))).*(0.3)./(1-
(0.3)))'),...
str2mat('0.0','0.0','0.0','0.0'),...
str2mat('0.0','0.0'),...
str2mat('1.0','0','0','1.0''),...
'0:10',...
'0.0',...
'0.0',...
'[0 100]')
setappdata(pde_fig,'currparam',...
['200E3';..
'0.3 ';..
'0.0 ';...
```

```
‘0.0 ’;...
‘1.0 ’])
% Solve parameters:
setappdata(pde_fig,‘solveparam’,...
str2mat(‘0’,‘1000’,‘10’,‘pdeadworst’,...
‘0.5’,‘longest’,‘0’,‘1e-4’,‘’,‘fixed’,‘inf’))
% Plotflags and user data strings:
setappdata(pde_fig,‘plotflags’,[11 1 1 1 1 1 1 1 0 1 0 1 0 1 0 0 0 1]);
setappdata(pde_fig,‘colstring’,‘’);
setappdata(pde_fig,‘arrowstring’,‘’);
setappdata(pde_fig,‘deformstring’,‘’);
setappdata(pde_fig,‘heightstring’,‘’);
% Solve PDE:
pdetool(‘solve’)
```

EDGECRACKPLANESTRAIN.m
```
function pdemodel
[pde_fig,ax]=pdeinit;
pdetool(‘appl_cb’,4);
set(ax,‘DataAspectRatio’,[1 1.2000000000000002 1]);
set(ax,‘PlotBoxAspectRatio’,[1.25 0.83333333333333326 5]);
set(ax,‘XLim’,[0 0.5]);
set(ax,‘YLim’,[0 0.40000000000000002]);
set(ax,‘XTickMode’,‘auto’);
set(ax,‘YTickMode’,‘auto’);
pdetool(‘gridon’,‘on’);
% Geometry description:
pdeellip(-
0.25877862595419843,0.41450381679389359,
   0.011450381679389166,0.066412213740458026,...
0,‘E1’);
pderect([0.0026717557251908775 0.30190839694656491
   0.12122137404580148
0.003053435114503789],‘R1’);
pdeellip(0.15152671755725192,0.1193893129770992,
   0.0015267175572519225,0.016488549618320622,...
0,‘E2’);
set(findobj(get(pde_fig,‘Children’),‘Tag’,‘PDEEval’),‘String’,‘R1-E2’)
% Boundary conditions:
pdetool(‘changemode’,0)
pdesetbd(8,...
‘neu’,...
2,...
str2mat(‘0’,‘0’,‘0’,‘0’),...
str2mat(‘0’,‘0’))
```

```
pdesetbd(7,...
'neu',...
2,...
str2mat('0','0','0','0'),...
str2mat('0','0'))
pdesetbd(6,...
'neu',...
2,...
str2mat('0','0','0','0'),...
str2mat('0','0'))
pdesetbd(5,...
'dir',...
2,...
str2mat('1','0','0','1'),...
str2mat('0','0'))
pdesetbd(4,...
'neu',...
2,...
str2mat('0','0','0','0'),...
str2mat('0','0'))
pdesetbd(3,...
'neu',...
2,...
str2mat('0','0','0','0'),...
str2mat('0','0'))
pdesetbd(2,...
'dir',...
2,...
str2mat('1','0','0','1'),...
str2mat('0','0'))
pdesetbd(1,...
'neu',...
2,...
str2mat('0','0','0','0'),...
str2mat('0.4','0'))
% Mesh generation:
setappdata(pde_fig,'Hgrad',1.3);
setappdata(pde_fig,'refinemethod','regular');
pdetool('initmesh')
% PDE coefficients:
pdeseteq(1,...
str2mat('2*((200E3)./(2*(1+(0.3))))+(2*((200E3)./(2*(1+(0.3))))).*(0.3)./(1-
2*(0.3)))','0','(200E3)./(2*(1+(0.3)))','0','(200E3)./
  (2*(1+(0.3)))','2*((200E3)./(2*(1+(0.3)))).*(0.3)./(1-
```

```
2*(0.3))’,‘0’,’(200E3)./(2*(1+(0.3)))’,‘0’,‘2*((200E3)./
    (2*(1+(0.3))))+(2*((200E3)./(2*(1+(0.3))))).*(0.3)./(1-
2*(0.3)))’),..
str2mat(‘0.0’,‘0.0’,‘0.0’,‘0.0’),...
str2mat(‘0.0’,‘0.0’),...
str2mat(‘1’,‘0’,‘0’,‘1’),...
‘0:10’,...
‘0.0’,...
‘0.0’,...
‘[0 100]’)
setappdata(pde_fig,‘currparam’,...
[‘200E3’;...
‘0.3 ’;...
‘0.0 ’;...
‘0.0 ’;...
‘1 ’])
% Solve parameters:
setappdata(pde_fig,‘solveparam’,...
str2mat(‘0’,‘1000’,‘10’,‘pdeadworst’,...
‘0.5’,‘longest’,‘0’,‘1e-4’,’’,‘fixed’,‘inf’))
% Plotflags and user data strings:
setappdata(pde_fig,‘plotflags’,[11 1 1 1 1 1 1 1 0 1 0 1 0 1 0 0 0 1]);
setappdata(pde_fig,‘colstring’,’’);
setappdata(pde_fig,‘arrowstring’,’’);
setappdata(pde_fig,‘deformstring’,’’);
setappdata(pde_fig,‘heightstring’,’’);
% Solve PDE:
pdetool(‘solve’)
```

MIX
Papier aus verantwortungsvollen Quellen
Paper from responsible sources
FSC® C105338

If you have any concerns about our products,
you can contact us on
ProductSafety@springernature.com

In case Publisher is established outside the EU,
the EU authorized representative is:
**Springer Nature Customer Service Center GmbH
Europaplatz 3, 69115 Heidelberg, Germany**

Printed by Libri Plureos GmbH
in Hamburg, Germany